Rhythm as Medicine, Rhythm as Illness

Why Sleep, Mood, Metabolism, and Immunity Rise and Fall Together

Moawiah Naffaa, PhD

Copyright and Disclaimer

This book presents a conceptual framework for understanding health and illness through the lens of biological rhythm, including sleep, mood, inflammation, metabolism, stress physiology, and immune function.

The **Oscillatory Health Model** offers a framework for exploring how biological timing and rhythmic regulation may relate to health and illness. It is intended to complement, rather than replace, established medical understanding and appropriate professional care.

Every effort has been made to ensure the accuracy and clarity of the scientific information presented at the time of publication.

Published by: Moawiah Naffaa

For inquiries or permissions:
Moawiah.Naffaa@proton.me

Table of Contents

PROLOGUE

The Day the Rhythm Broke

It did not happen all at once.

At first, it was only the sleep.

He began waking at 3:17 a.m.
Not in panic. Not from a dream.
Just awake — as if something inside him had forgotten how to descend.

The room was quiet. The body was tired. But the mind hovered.

The next morning he functioned. Coffee compensated. The day moved forward. No crisis.

But the following night it happened again.

And then again.

Within a week, something subtle had shifted. Not dramatic enough to alarm anyone — not even him — but enough to alter the texture of the day.

His mood flattened.

Not sadness exactly. Not despair. Just a thinning. The sharp edges of enthusiasm dulled. Conversations required more effort. Music felt distant. Small pleasures registered, but faintly — as though filtered through glass.

He told himself he was simply tired.

Then came the cold that lingered longer than usual.
A sore throat that refused to resolve.
A faint ache in the joints.
A low-grade inflammation of being.

He dismissed that too.

Work grew heavier. Meals felt less satisfying. He craved sugar in the afternoon and found himself restless at night. His energy no longer followed a clean arc from morning to evening. It staggered. Spiked. Crashed.

He began declining invitations.
He answered messages later.
He canceled plans he would once have kept.

Nothing catastrophic had happened.

No diagnosis.
No trauma.
No singular event.

And yet something was clearly wrong.

It felt as though several small problems were occurring at once:

Poor sleep.
Lower mood.
Weaker immunity.
Unstable energy.
Less social interest.

Five separate issues.

Five different systems.

Five different explanations.

But beneath them all, there was only one change.

The rhythm had broken.

The Illusion of Separate Systems

We are taught to think of the body in compartments.

Sleep is one system.
Mood is another.
Immunity another.
Metabolism another.
Relationships something else entirely.

So when one falters, we look for a localized cause.

But the body does not operate in compartments. It operates in waves.

Sleep is a rhythm.
Cortisol is a rhythm.
Insulin is a rhythm.
Cytokines rise and fall.
Neurotransmitters pulse.
Heart rate oscillates.
Even attention moves in cycles.

Health is not a static condition. It is a coordinated choreography of oscillations — rising and falling, activating and recovering, tightening and releasing.

When one wave loses timing, others follow.

In this case, sleep lost its phase.
Recovery weakened.
Cortisol shifted upward.
Inflammation rose.
Dopamine amplitude narrowed.
Energy desynchronized.

Nothing "broke" in isolation.

The orchestra fell out of time.

The Cost of Lost Oscillation

When rhythms flatten, the body does not immediately collapse.

It stiffens.

The immune system no longer pulses efficiently between activation and resolution.
Metabolism no longer anticipates light and food.
The brain no longer alternates cleanly between focus and rest.

Recovery shortens.
Baseline drifts.
Amplitude narrows.

What looks like fatigue is often reduced oscillatory range.
What feels like low mood may be dampened amplitude.
What presents as chronic inflammation may be a baseline that no longer resets.

The body becomes less flexible.

And flexibility is often central to health.

The Hidden Grammar of Wellness

This book begins with a simple but radical premise:

Health does not depend on static stability.

It often depends on flexible biological oscillation.

Every major physiological system — neural, endocrine, immune, metabolic, social — can be understood through the properties of its rhythmic movement.

Amplitude.
Frequency.
Phase.
Synchronization.
Recovery.
Baseline.

When these variables remain flexible and coordinated, we experience vitality.

When one or more distort persistently, symptoms emerge.

Not as moral failures.
Not as personal weakness.
But as pattern distortion.

Nothing "Separate" Broke

The man who could not sleep did not suddenly develop five different problems.

He lost rhythm.

The body's cycles fell slightly out of alignment.
Small misalignments compounded.
Oscillations narrowed.
Recovery shortened.
Baselines drifted.

The system did what all systems do when flexibility declines — it became rigid.

And rigidity, over time, becomes illness.

The Way Back

The remarkable truth is this:

Rhythm can be restored.

The nervous system can relearn amplitude.
The immune system can regain pulsatility.
Metabolism can re-synchronize.
Sleep can re-anchor to light.
Mood can regain range.

Not instantly.
Not mechanically.
But biologically.

Because oscillation is the native language of life.

We do not lose health all at once.

We lose oscillation.

And if rhythm is the problem,
rhythm can also be the medicine.

PART I — The Oscillating Organism

The Architecture Beneath Health

Before we can speak of illness, we must understand design.

Long before medicine gave names to disorders, long before laboratories measured hormones or genes, life was already moving in waves. The human body did not begin as a machine with parts; it began as a rhythm. Cells divide in cycles. The heart contracts and releases. Neurons fire in pulses. Hormones rise and fall. Even breath is an oscillation — inhale, exhale — a quiet reminder that existence itself depends on alternation.

We often imagine health as stillness — a calm, stable plateau where nothing fluctuates too much. But the body does not aim for flatness. It aims for movement within bounds. A healthy heart does not beat like a metronome; it varies subtly from one beat to the next. A resilient immune system does not remain silent; it activates and resolves. A balanced mind does not feel one emotion forever; it expands and contracts, rises and settles.

The deeper we look into biology, the clearer this becomes: life is not static equilibrium. It is dynamic regulation.

This part of the book asks a simple but transformative question:

What if oscillation is not a feature of health — but its foundation?

When we begin to observe the body through this lens, the familiar categories of sleep, mood, metabolism, immunity, and connection start to blur. They are not independent systems operating in isolation. They are rhythmic processes moving in relation to one another. Each system has its own tempo, its own amplitude, its own timing. Health emerges not from the perfection of any single rhythm, but from the harmony between them.

A symphony does not require every instrument to play the same note. It requires coordination.

The oscillating organism is not chaotic. It is patterned. Waves follow laws. Rhythms obey structure. Behind the visible fluctuations of energy and emotion lies a hidden mathematical grammar — a set of principles that govern how biological systems rise, fall, synchronize, and recover.

To understand disease, we must first understand this grammar.
To restore health, we must first recognize the rhythm.

In the chapters that follow, we will move from observation to structure. We will examine the subtle oscillations already occurring inside you — in your heart, your brain, your hormones, your immune cells — and begin to see them not as isolated signals, but as coordinated waves.

We will discover that what feels like stability is actually flexibility. That what appears calm is often dynamic balance. That vitality is not the absence of fluctuation, but the ability to move through it without losing coherence.

This is the architecture beneath health.

Not rigid.
Not linear.
Not mechanical.

But oscillatory.

Life does not hold one note.

It breathes between them.

Chapter 1 — Life Moves in Waves

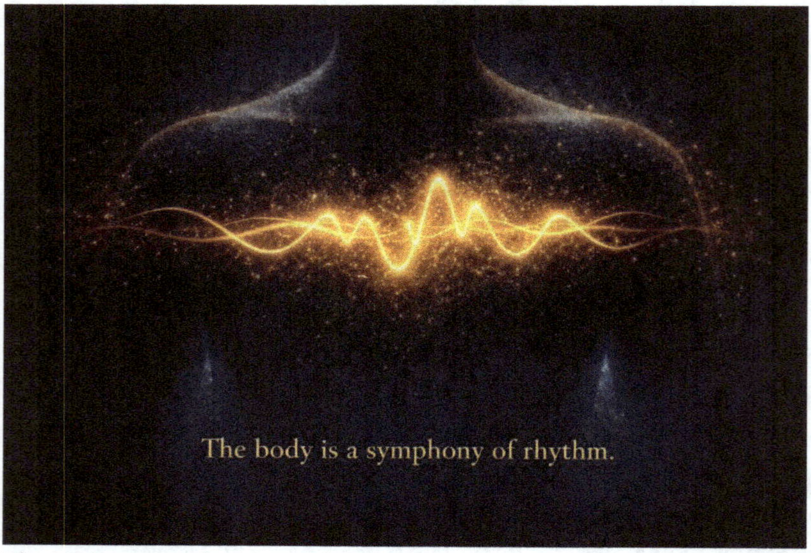

If you place your fingers gently against your wrist, you will feel a rhythm.

It seems steady at first — beat after beat, predictable and reassuring. But if you could magnify that rhythm and slow it down, you would notice something remarkable. The intervals are not identical. One beat arrives a fraction sooner, another a fraction later. The heart does not tick like a clock. It breathes.

This subtle variation is called heart rate variability, and it is one of the clearest signatures of vitality. A heart that varies slightly from moment to moment is not unstable — it is adaptable. It is listening to the nervous system, adjusting to breath, posture, thought, and emotion. Its micro-fluctuations reflect flexibility. When that variability disappears and the rhythm becomes rigidly uniform, health declines.

The heart is not alone.

Hormones pulse in waves rather than flowing in constant streams. Cortisol rises in the early morning, preparing the body for wakefulness, then gradually declines. Insulin surges in response to food and falls again. Growth hormone releases in bursts during deep sleep. Even reproductive hormones follow monthly oscillations, weaving cycles of fertility and rest.

The brain, too, is rhythmic. Neurons fire in synchronized patterns known as neural oscillations — alpha waves during relaxed awareness, beta waves during focused thought, delta waves in deep sleep. These electrical rhythms coordinate perception, memory, and emotion. When neural timing falters, attention fragments and mood destabilizes.

The immune system, often imagined as a silent defense force, is also cyclical. Immune cells patrol more actively at certain times of day. Cytokines — the chemical messengers of inflammation — rise and fall in regulated patterns. Fever itself is not a malfunction but a controlled oscillatory response, a temporary elevation designed to reset balance.

From the heartbeat to the sleep cycle, from glucose regulation to emotional tone, life unfolds in waves.

Yet we rarely think of it this way.

We speak of "normal levels" as though the body were meant to hold a constant line. We measure averages and baselines. We seek stability as if it were stillness. But biology does not aim for flatness. It aims for dynamic equilibrium — movement within bounds.

This is the difference between homeostasis and dynamic regulation.

Homeostasis is often described as the maintenance of internal stability. But true biological stability is not rigid sameness. It is controlled fluctuation. The nervous system constantly adjusts heart rate in response to breathing. Hormones shift minute by minute depending on light, food, stress, and social interaction. The immune system activates and resolves in cycles. The body remains "stable" not because it avoids change, but because it moves through change smoothly.

Fluctuation is not the enemy of health. It is often one of its core mechanisms.

When waves are allowed to rise and fall within adaptive ranges, systems remain responsive. They can accelerate when needed and recover when danger passes. They can engage and disengage. They can expand and contract.

But when oscillation narrows or disappears, rigidity sets in.

A heart that no longer varies becomes fragile.
An immune system that cannot quiet itself becomes chronically inflamed.
A stress response that does not subside becomes exhaustion.
A mood that cannot lift becomes depression.
A metabolism that cannot flex becomes dysregulated.

The loss of wave is the beginning of stiffness.
The loss of stiffness is the beginning of collapse.

Health depends not on preventing change, but on preserving adaptability.

To understand this more deeply, imagine the body not as a collection of independent systems but as a living rhythm — a network of oscillators communicating across time. The heart responds to breath. Breath responds to emotion. Emotion responds to social cues. Hormones respond to light. Immune cells respond to sleep. Each rhythm influences the next.

These oscillations do not operate in isolation. They are nested within one another like waves within waves — seconds within hours, hours within days, days within seasons. The daily circadian rhythm coordinates smaller pulses. Neural oscillations align with hormonal patterns. The body behaves less like a machine with parts and more like an orchestra with instruments.

No instrument sustains a single note forever.
Music exists because of variation.

In the same way, vitality emerges from coordinated fluctuation.

As we move through this chapter, we will look closely at these hidden oscillations — in the heart, the endocrine system, the brain, and the immune network. We will begin to see that what we call "normal" is not a fixed point, but a dynamic range. And we will recognize that the capacity to move — to rise, fall, and recover — is the defining feature of biological life.

Life is not built on straight lines.

Life moves in waves.

And as we will discover, across biological systems, life consistently expresses oscillatory design.

1.1 The Hidden Oscillation

Beneath the appearance of steadiness, the body is in constant motion.

Not dramatic motion. Not visible motion.
But microscopic, rhythmic fluctuation — a continuous rise and fall occurring in nearly every system at once.

Take the heart.

To the ear, it sounds regular: lub-dub, lub-dub, lub-dub. Yet if you measure the milliseconds between beats, you discover that they subtly change from one beat to the next. This variation is called **heart rate variability**. When you inhale, the heart speeds slightly. When you exhale, it slows. When you relax, variability increases. When you are chronically stressed, it narrows.

A perfectly uniform heartbeat is not a sign of strength. It is a sign of rigidity.

Healthy cardiac function depends on this micro-oscillation — a constant dialogue between the sympathetic and parasympathetic nervous systems. The heart is not just pumping; it is listening. It adjusts moment by moment to internal and external cues. Its variability reflects adaptability.

The endocrine system behaves similarly.

Hormones are not secreted at fixed levels. They are released in pulses. Cortisol surges in the early morning, helping you wake. It gradually declines throughout the day. Insulin spikes after meals and recedes as glucose stabilizes. Growth hormone is released in bursts during deep sleep. Even reproductive hormones follow rhythmic monthly patterns.

If hormones were constant, they would lose meaning. Pulsatility allows responsiveness. A surge signals change; a decline allows recovery.

The brain, too, speaks in oscillations.

Neurons fire in synchronized electrical patterns known as **neural rhythms**. During focused attention, faster beta waves dominate. During relaxed wakefulness, alpha waves rise. In deep sleep, slow delta waves sweep across the cortex. These oscillations coordinate perception, memory consolidation, emotional processing, and motor control.

When neural rhythms become chaotic or excessively rigid, cognition fragments. Attention drifts. Mood destabilizes. The brain requires coordinated oscillation to function coherently.

Even the immune system — often imagined as a constant defense barrier — operates in cycles.

Immune cells circulate more actively at certain times of day. Inflammatory signals rise during acute threat and fall during resolution. Cytokines follow circadian patterns, and the magnitude of immune response changes across the 24-hour cycle. Fever itself is a temporary oscillatory elevation — a controlled deviation meant to restore equilibrium.

What we call "defense" is in fact dynamic cycling between activation and quieting.

Across cardiovascular, endocrine, neural, and immune systems, the pattern repeats:

Rise.
Peak.
Decline.
Return.

Oscillation is not accidental. It is built into the architecture of life.

1.2 Why Stability Is Dynamic

We often describe health as "stable."

But what does stability actually mean?

If stability were stillness, a healthy heart would beat at identical intervals. Hormones would remain at fixed levels. Brain activity would hold one frequency. Immune signals would neither rise nor fall.

Such stillness would not be life. It would be failure.

Biology achieves stability not through immobility, but through controlled fluctuation.

This distinction marks the difference between traditional notions of **homeostasis** and the more accurate concept of **dynamic regulation**.

Homeostasis is commonly defined as the maintenance of internal constancy — body temperature near 37°C, blood glucose within a narrow range, pH tightly regulated. But how is this constancy achieved? Not by freezing variables in place, but by continuously adjusting them.

Blood glucose rises after eating and falls between meals. Temperature fluctuates slightly across the day. Blood pressure shifts with posture and emotion. These small deviations are not errors; they are adjustments.

Dynamic regulation means the system is constantly correcting itself — moving within safe ranges rather than clinging to a single point.

Imagine standing upright. You do not remain perfectly still. Tiny muscle contractions continually adjust your balance. Without those micro-movements, you would fall. Stability requires motion.

The same principle governs physiology.

Fluctuation is normal because the environment is not static. Light changes. Food intake changes. Social interactions shift. Stressors appear and disappear. A system that could not oscillate would be unable to respond.

Adaptability depends on variability.

When oscillations are flexible, the body can:

- Accelerate when needed.
- Recover when danger passes.
- Synchronize with day–night cycles.
- Shift between focus and rest.
- Mount and resolve immune responses.

Stability emerges from responsiveness, not rigidity.

This is why reduced variability often signals vulnerability. A heart that cannot vary may struggle under stress. A stress response that cannot subside becomes chronic. An immune system that cannot quiet itself generates persistent inflammation.

The healthiest systems are not those that fluctuate least, but those that fluctuate within adaptive ranges.

Fluctuation is not noise.
It is information.

And as we will see, when that information becomes distorted — when oscillations narrow, accelerate, desynchronize, or fail to recover — health begins to erode.

Life maintains balance by moving.

Stillness in biology is rarely a sign of strength.

It is motion, carefully regulated, that keeps us alive.

1.3 The Cost of Flattening

If oscillation is the signature of life, then flattening is the first whisper of decline.

Waves are meant to rise and fall. When they stop rising, when they stop falling, when variation narrows into a thin, rigid line, something essential is lost: adaptability.

Consider the heart again. A healthy heart varies subtly with each breath, adjusting to internal and external demands. When variability disappears and beats become mechanically uniform, the system is less responsive to stress. The capacity to accelerate when needed or to slow when safe diminishes. Rigidity replaces flexibility.

The same pattern appears elsewhere.

A mood that never lifts, never dips — emotionally flat — loses its range. Joy becomes muted. Motivation weakens. Recovery slows. Emotional adaptability shrinks.

An immune system that cannot quiet itself remains chronically elevated. Inflammation persists even when no threat remains. What once protected now damages.

A stress response that fails to resolve becomes constant vigilance. Cortisol no longer rises and falls cleanly across the day. It lingers, eroding sleep, metabolism, and mood.

Flattening is not always dramatic. It is often subtle — a narrowing of amplitude, a shortening of recovery, a slight upward drift in baseline. But over time, these small distortions accumulate.

Rigidity feels stable at first. There is predictability in it. A flat line does not surprise. But rigidity sacrifices responsiveness. It reduces the system's ability to adapt to new demands.

Adaptability, not stillness, defines resilience.

A flexible system can expand under challenge and contract afterward. It can mount an immune response and then resolve it. It can experience intense emotion and then return to baseline. It can exert effort and then recover.

When waves disappear, so does that flexibility.

Biological rigidity eventually becomes fragility.

And fragility, under stress, becomes illness.

1.4 The Body as a Living Rhythm

Up to this point, we have examined oscillations system by system — heart, hormones, brain, immunity. But the body does not function as a set of isolated waves.

It functions as a coordinated field of rhythms.

The heartbeat aligns subtly with breath.
Cortisol rhythms influence immune activity.
Sleep timing shapes metabolic responses.
Neural oscillations synchronize during social interaction.

These are not coincidences. They are forms of entrainment — the natural tendency of oscillators to align when connected.

In physics, when two pendulums hang from the same beam, they gradually synchronize. In biology, when systems share signals — neural, hormonal, metabolic — they begin to move together.

The body is a network of oscillators nested within one another.

Seconds within minutes.
Minutes within hours.
Hours within days.
Days within seasons.

Circadian rhythms coordinate hormonal pulses. Hormonal pulses influence neural firing. Neural rhythms shape behavior. Behavior feeds back into physiology.

Health emerges not from perfect individual rhythms, but from coordinated rhythms.

If one oscillator drifts slightly, others compensate. But if drift persists, synchronization weakens. The orchestra begins to lose tempo.

The body is not a machine with separate parts operating independently. Machines require external calibration. The body recalibrates internally through feedback loops, constant adjustment, and rhythmic communication.

It is less like a clock and more like music.

Music requires variation. Crescendos and decrescendos. Tension and release. Silence and sound.

Without variation, there is no melody.

Without coordination, there is noise.

The living organism maintains health by sustaining rhythmic coherence across systems. This coherence allows flexibility without chaos, responsiveness without collapse.

When rhythms align, we feel energetic yet calm, focused yet relaxed, engaged yet capable of rest. When rhythms desynchronize, we feel fatigued, irritable, inflamed, or disconnected — even if no single organ appears "broken."

This is the deeper shift in perspective this book invites.

We do not simply possess organs.

We embody oscillations.

And when we step back far enough to see the whole pattern, one truth becomes clear:

Life is not static.

Life is oscillatory by design.

Chapter 2 — The Mathematics of Health

The Hidden Structure Beneath the Waves

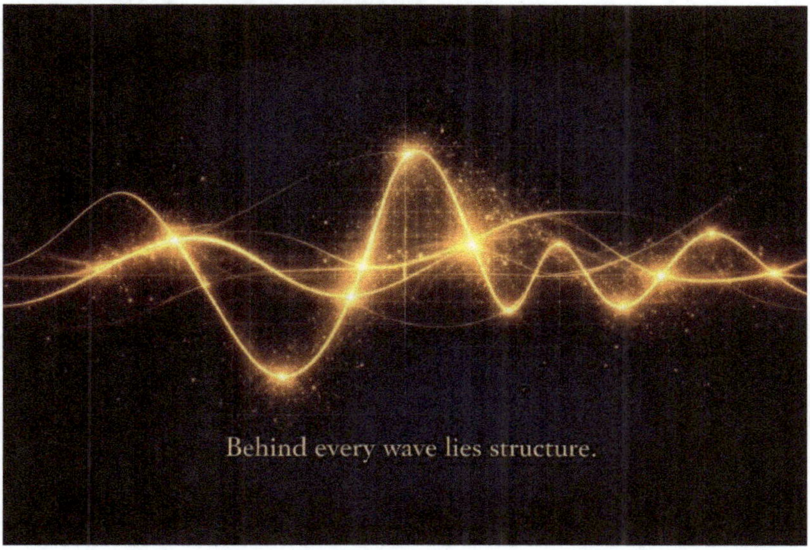

In the previous chapter, we established a central observation: life moves in waves.

This chapter formalizes that observation.

Observation alone is not enough.
If oscillation underlies health, we must define its structure.
What makes one wave resilient and another distorted?
What variables determine whether rhythm protects or erodes vitality?

Every oscillatory system can be described by a small set of defining properties.

Whether acoustic, electromagnetic, or biological, waves follow structural rules.

Biological rhythms are no exception.

Behind the visible rise and fall of heart rate, cortisol, mood, glucose, and immune activity lies a structural grammar — a small set of variables that determine how those rhythms behave over time.

Without this grammar, we are left describing symptoms.
With it, we can describe patterns.

Observing that someone sleeps poorly tells us little.
Understanding whether the problem is amplitude, phase, or recovery tells us more.

Seeing inflammation rise is informative.
Understanding whether the baseline has shifted is transformative.

In medicine, we often focus on levels — high or low, elevated or deficient. But levels are snapshots. Health unfolds over time. It is a dynamic pattern, not a static measurement.

To understand dynamic systems, we need variables that describe movement.

This chapter introduces what I will call the **Oscillatory Health Model** — a framework built on six fundamental variables that define every biological rhythm:

- Amplitude
- Frequency
- Phase
- Synchronization
- Recovery
- Baseline

Together, they describe how strong a wave is, how fast it moves, whether it is aligned with its environment, whether it coordinates with other waves, how efficiently it returns to rest, and where that resting point lies.

These six variables are not abstract mathematics. They are living properties of your nervous system, your endocrine system, your immune function, your metabolism, and even your emotional life.

They allow us to translate experience into structure.

They allow us to see that:

Depression may reflect flattened amplitude.
Anxiety may reflect accelerated frequency.
Insomnia may reflect phase disruption.
Chronic inflammation may reflect baseline shift.
Burnout may reflect desynchronization and impaired recovery.

Not as simplistic explanations — but as structural patterns.

Once we have variables, we can map distortion.
Once we can map distortion, we can restore flexibility.

The Oscillatory Health Model provides a structural grammar for biological regulation.
In the chapters that follow, we will apply these six variables across sleep, mood, immunity, metabolism, stress, and social regulation.

The question is no longer whether rhythm exists.
The question is how it is behaving.

2.1 Why We Need a Framework

Recognition without structure remains metaphor.

If oscillation is foundational to health, it must be measurable, describable, and operational.

A framework transforms rhythm from poetry into physiology.

To say that "everything moves in cycles" is true — yet too vague to guide understanding. If we stop there, rhythm becomes metaphor. Beautiful, but imprecise.

Medicine does not progress through metaphors.
It progresses through structure.

If we want to understand why one person sleeps deeply and another lies awake at 3 a.m., if we want to understand why inflammation resolves in one body but persists in another, if we want to understand why some moods recover and others remain flattened — we need more than the word *cycle*.

We need variables.

Imagine watching the ocean.

You see waves rising and falling. But to truly understand them, you would ask:

How high are they?
How often do they arrive?
Are they aligned with the tide?
Do they reinforce one another or collide?

How quickly do they settle after a storm?
Where does the water rest between surges?

Without those distinctions, all waves look the same.

Biology is no different.

Two people may both experience stress. In one, cortisol rises sharply and then falls cleanly — strong amplitude, efficient recovery. In another, cortisol rises only slightly but never fully declines — narrow amplitude, impaired recovery. Outwardly similar. Structurally different.

One person may feel anxious because their internal rhythms accelerate — frequency distortion. Another may feel exhausted because their rhythms flatten — amplitude reduction. Same category. Different pattern.

Observing that "something is off" does not tell us *how* it is off.

Variables do.

A framework allows us to move from surface description to structural explanation.

It allows us to ask:

Is the problem the strength of the rise?
The speed of the cycle?
The timing relative to light and environment?
The coordination between systems?
The ability to return to baseline?
Or the baseline itself?

Without variables, we treat symptoms in isolation.

With variables, we see patterns across systems.

This is the crucial shift.

Sleep problems are not separate from mood problems if both reflect phase misalignment.
Metabolic instability is not separate from stress dysregulation if both reflect desynchronization.
Chronic inflammation is not separate from emotional flattening if both reflect baseline shift.

A framework reveals unity beneath diversity.

It does not oversimplify disease. It organizes it.

In complex systems — whether climate, music, or physiology — patterns emerge from a small set of governing properties. When those properties remain flexible, the system adapts. When they become distorted, instability spreads.

Health is no exception.

To understand oscillation deeply, we must describe its architecture.

And that architecture rests on six variables.

They are simple.
They are universal.
And once you see them, you will begin to recognize them everywhere — in your sleep, your energy, your mood, your stress response, your immune resilience.

Observing waves is awareness.

Understanding their variables is insight.

The next section will introduce the grammar of biological rhythm — the six properties that define how every living wave behaves.

2.2 The Six Variables of Biological Rhythm

Every oscillatory system can be mathematically described using a small set of variables.

Biological rhythms are no exception.
Imagine a simple sine wave drawn across a page.

It rises.
It peaks.
It falls.
It returns.

Now imagine adjusting that wave.

Make it taller.
Make it faster.
Shift it to the right.
Layer another wave on top.

Delay its return.
Move its resting line upward.

Without changing its nature as a wave, you have completely changed its behavior.

Biology works the same way.

Every heartbeat, every cortisol cycle, every mood fluctuation, every immune response can be described using six fundamental variables:

- **Amplitude**
- **Frequency**
- **Phase**
- **Synchronization**
- **Recovery**
- **Baseline**

These are not abstract mathematical ideas. They are living properties of your physiology.

Each variable answers a specific question about how a rhythm behaves.

Let us take them one by one.

2.2.1 Amplitude

How strong is the rise and fall?

Definition
Amplitude refers to the magnitude of change between peak and trough — how large the oscillation is relative to baseline.

Visual Metaphor
Picture ocean waves. Some barely ripple across the surface. Others crest high before falling back. The difference between a small ripple and a powerful surge is amplitude.

Biological Example
A healthy dopamine response rises clearly in anticipation of reward and declines afterward. A healthy immune response mounts decisively against infection and resolves once the threat is gone. A healthy cortisol rhythm peaks strongly in the morning and falls by night.

Structural Importance
Amplitude determines responsiveness. Without sufficient rise, the system

cannot mobilize. Without sufficient fall, it cannot rest. Amplitude gives life its contrast — effort followed by release, activation followed by calm.

Hint of Distortion

When amplitude flattens, energy dulls, emotion narrows, and motivation fades. When amplitude becomes excessive, volatility and instability emerge.

2.2.2 Frequency

How often does the cycle occur?

Definition

Frequency refers to the rate at which oscillations repeat over time.

Visual Metaphor

Imagine waves hitting the shore. Some arrive slowly and rhythmically. Others crash in rapid succession. The number of waves per minute is frequency.

Biological Example

Breathing follows a frequency. Heartbeats follow a frequency. Stress responses can cycle slowly or rapidly. Anxiety often reflects accelerated internal cycling — rapid pulses of arousal without sufficient spacing.

Structural Importance

Frequency governs tempo. Systems must cycle often enough to respond but slowly enough to recover. Too slow and responsiveness weakens. Too fast and recovery becomes impossible.

Hint of Distortion

When frequency accelerates excessively, the body feels restless, hypervigilant, unable to settle. When it slows too much, lethargy and stagnation appear.

2.2.3 Phase

Is timing aligned with environmental cues?

Definition

Phase refers to the temporal alignment of a rhythm relative to external signals — especially light–dark cycles, food timing, and social activity.

Visual Metaphor

Imagine two pendulums swinging. If one lags slightly behind the other, they are out of phase. Timing matters as much as magnitude.

Biological Example

Cortisol should rise in the morning. Melatonin should rise at night. Insulin sensitivity follows circadian timing. When sleep shifts later and light exposure changes, internal rhythms drift out of phase.

Structural Importance

Phase ensures that internal rhythms align with the environment. The body anticipates light, food, activity, and rest. Correct timing conserves energy and enhances efficiency.

Hint of Distortion

Phase delay contributes to insomnia, metabolic disruption, and mood instability. Misaligned timing gradually desynchronizes entire systems.

2.2.4 Synchronization

Do systems rise and fall together?

Definition

Synchronization refers to the coordination between multiple oscillators — whether rhythms align and reinforce one another.

Visual Metaphor

In an orchestra, instruments must enter together. If one speeds ahead while others lag, harmony dissolves.

Biological Example

Sleep cycles coordinate with immune activity. Hormonal pulses influence neural rhythms. Heart rate variability aligns with breathing. Social interaction can synchronize nervous systems through co-regulation.

Structural Importance

Health depends on cross-system coherence. Individual rhythms may function well alone, but without coordination the organism loses efficiency. Synchronization allows energy to flow smoothly across systems.

Hint of Distortion

Desynchronization manifests as burnout, metabolic syndrome, chronic fatigue, and multi-system instability — not because one organ failed, but because timing between systems fractured.

2.2.5 Recovery

How efficiently does the system return to baseline?

Definition
Recovery refers to the speed and completeness with which a rhythm returns to its resting state after activation.

Visual Metaphor
After a wave crashes, how quickly does the water settle? Does it retreat cleanly, or remain turbulent?

Biological Example
After acute stress, cortisol should fall. After infection, inflammation should resolve. After exertion, heart rate should normalize. After sadness, mood should gradually lift.

Structural Importance
Recovery protects against chronic activation. It preserves flexibility. Without adequate recovery, the next cycle begins from an elevated state.

Hint of Distortion
Impaired recovery contributes to chronic stress, persistent inflammation, exhaustion, and emotional depletion.

2.2.6 Baseline

Where does the system rest between cycles?

Definition
Baseline is the resting set-point around which oscillations occur.

Visual Metaphor
Imagine lifting the entire wave upward or downward without changing its shape. The resting line has shifted.

Biological Example
Chronically elevated cortisol reflects a raised baseline. Persistent low mood reflects a lowered affective baseline. Chronic inflammation reflects an upward shift in immune resting tone.

Structural Importance
Baseline determines starting position. Even healthy oscillations become problematic if the resting point drifts too high or too low. Small shifts accumulate across systems.

Hint of Distortion
Baseline shifts are often subtle but powerful. They may precede overt disease by months or years.

These six variables form the grammar of biological rhythm.

Every oscillating system in the body can be described in their terms.
Every distortion of health can be mapped across one or more of them.

They do not replace traditional diagnoses.
They organize them.

And together they lead to a simple but profound principle:

Health is not the absence of fluctuation.

Health is the flexible modulation of amplitude, frequency, phase, synchronization, recovery, and baseline.

2.3 The Oscillatory Health Matrix

We now have the six variables.

But understanding them individually is not enough.
We must see how they function **together**.

Biology does not distort in one dimension at a time. A shift in **phase** can affect **synchronization**. Impaired **recovery** can gradually raise **baseline**. Altered **frequency** can compress **amplitude**. The system is interdependent.

To make this complexity usable, we need a clear structural summary.

The **Oscillatory Health Matrix** is that summary.

It allows us to step back from symptoms and ask a deeper question:

Which **variable** is being distorted?

Below is a simplified conceptual map of the model.

The Oscillatory Health Matrix

Variable	Healthy Expression	Distorted Pattern	What It Feels Like
Amplitude	Clear, adaptive rise and fall	Flattened or exaggerated response	Emotional blunting or volatility
Frequency	Sustainable tempo of cycles	Too rapid or too slow cycling	Restlessness or lethargy

Variable	Healthy Expression	Distorted Pattern	What It Feels Like
Phase	Proper alignment with light, food, and activity	Timing misalignment	Insomnia, metabolic drift
Synchronization	Systems rise and fall in coordination	Cross-system desynchronization	Burnout, feeling "out of sync"
Recovery	Efficient return to resting state	Prolonged activation, incomplete resolution	Lingering stress, persistent inflammation
Baseline	Adaptive resting set-point	Elevated or depressed resting tone	Chronic tension or chronic fatigue

Notice what this matrix does.

It shifts the question from:

"What is wrong with me?"

to

"How is my rhythm behaving?"

Two people may both feel exhausted.
One may have distorted **phase**.
Another may have impaired **recovery**.
A third may have a shifted **baseline**.

The symptom is similar.
The structure is different.

The matrix also reveals something essential:

Health is not about eliminating fluctuation.
It is about preserving **flexibility**.

A healthy system can:

- Increase **amplitude** when needed
- Adjust **frequency** under stress
- Realign **phase** with environment
- Maintain **synchronization** across systems
- Complete **recovery** after activation
- Restore an adaptive **baseline**

When flexibility narrows, distortion accumulates.

When flexibility remains intact, resilience persists — even under strain.

This is the structural heart of the model.

Not perfection.
Not constant calm.
Not static balance.

But modulation.

And that leads us to the defining principle of this book:

Health = flexible modulation of amplitude, frequency, phase, synchronization, recovery, and baseline.

Figure 2.3 — The Oscillatory Health Model. *The six structural variables that define every biological rhythm. Health reflects flexible modulation across amplitude, frequency, phase, synchronization, recovery, and baseline.*

The Oscillatory Health Model

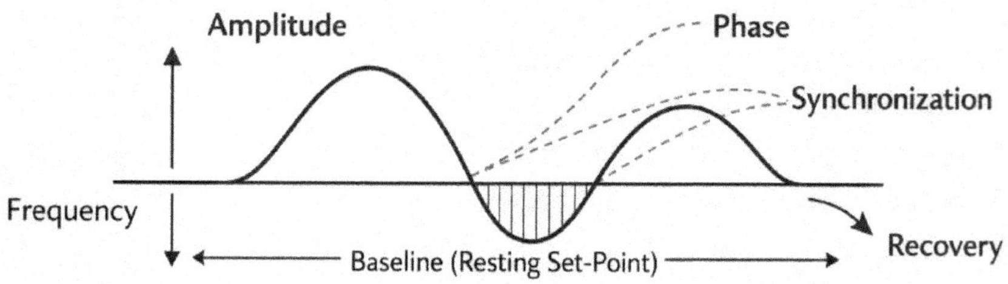

Variable	Healthy Pattern	Distortion Pattern
Amplitude	Peak	Flat
Frequency	Rhythm	Arrhythmic
Phase	Alignment	Mistimed
Synchronization	Coherence	Discordant
Recovery	Resolution	Lingering
Baseline	Flexibility	Rigidity

Health = Flexible Modulation of Six Variables.

Figure 2.3 The Oscillatory Health Model. *Biological rhythms can be described using six fundamental variables:* ***amplitude*** *(strength of rise and fall),* ***frequency*** *(rate of repetition),* ***phase*** *(timing relative to environmental cues),* ***synchronization*** *(coordination across systems),* ***recovery*** *(return to baseline after activation), and* ***baseline*** *(resting set-point). Distortion of one or more variables contributes to pattern-level dysregulation across sleep, mood, metabolism, immunity, and stress systems.*

This model will serve as a reference point throughout the remainder of the book. In the chapters that follow, we will identify which variable is primarily involved in different patterns of resilience and distortion, and how flexible modulation restores health.

Chapter 3 — From Neurotransmitters to Circadian Clocks

Chemistry as Rhythm

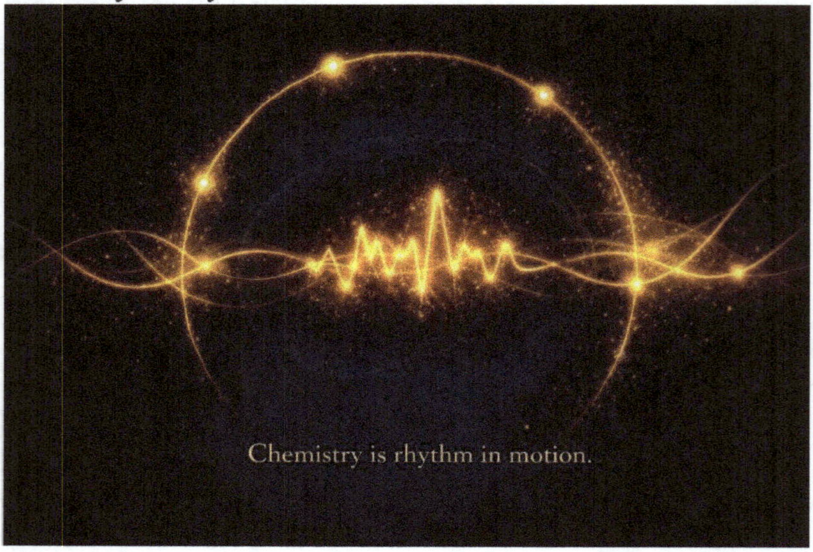

When we speak of dopamine, serotonin, cortisol, or melatonin, we often imagine substances — chemicals moving through blood or synapses, adjusting mood or metabolism like dials on a machine.

But this image is incomplete.

These molecules are not static substances waiting at fixed levels.
They are **temporal signals**.
They rise.
They fall.
They pulse.
They align with light, sleep, stress, anticipation, and social interaction.

They are not simply chemicals.

They are **oscillators**.

Dopamine does not sit in the brain like fuel in a tank. It fires in bursts. Short pulses encode surprise, motivation, and learning. Without those pulses, the brain cannot distinguish between expectation and reality.

Serotonin does not simply "increase mood." It modulates rhythm. It shapes pacing. It stabilizes emotional amplitude over time, smoothing extremes and reinforcing continuity.

Cortisol is not merely a stress hormone. It follows a daily wave — rising before dawn, peaking in early morning, gradually declining into night. When this wave shifts, sleep fragments, immunity falters, and metabolism drifts.

Melatonin does not cause sleep in isolation. It signals darkness. It marks phase. It is the body's internal declaration that night has arrived.

And beneath them all lie clock genes — molecular oscillators embedded within nearly every cell. These genes turn on and off in 24-hour cycles, creating internal time. They coordinate metabolism, immunity, cognition, and repair.

From synapse to nucleus, from neurotransmitter pulse to genetic oscillation, biology is organized around timing.

The deeper we look, the more we see that chemistry is not static concentration. It is patterned fluctuation.

If dopamine were constant, motivation would disappear.
If cortisol never declined, recovery would fail.
If melatonin rose at noon, metabolism would desynchronize.
If clock genes stopped cycling, cellular coordination would collapse.

These signals derive their meaning not from their presence, but from their movement.

This chapter traces the journey from neurotransmitter bursts measured in milliseconds to circadian gene oscillations measured in hours and days. We will see how these rhythms nest within one another — fast pulses embedded inside slower waves, minute cycles within daily cycles.

We will begin to understand that mood, energy, sleep, stress, and immunity are not controlled by isolated chemicals. They are governed by layered oscillatory networks.

And once this perspective becomes clear, something shifts:

We stop asking, "Is my serotonin low?"
And begin asking, "How is my rhythm behaving?"

Because what we call chemicals are, in truth, carriers of time.

They do not merely change levels.

They move.

And life, at every scale, moves with them.

3.1 Dopamine Pulses

Dopamine is often described as the molecule of pleasure.

But pleasure is not its primary language.

Dopamine speaks in **bursts**.

When something unexpected happens — a reward arrives, a prediction is confirmed, a possibility appears — dopamine neurons fire in rapid pulses. These pulses last milliseconds. They encode **change**, not contentment. They signal that something in the environment is different from what was anticipated.

Between bursts, dopamine activity quiets.

This pulsing pattern is essential. If dopamine were constant, nothing would stand out. Motivation would flatten. Learning would stall. The brain depends on contrast — on the rise and fall of signal — to distinguish significance from background noise.

Dopamine's meaning lies in its oscillation.

A healthy dopamine system rises sharply in anticipation and falls cleanly afterward. The amplitude gives energy. The decline allows recovery. When amplitude narrows, motivation dulls. When pulses become erratic or exaggerated, behavior becomes impulsive.

Dopamine is not a reservoir of pleasure.

It is a rhythmic teaching signal — a fast oscillator shaping desire, effort, and learning.

3.2 Serotonin Stabilization

If dopamine is the spark, serotonin is the stabilizer.

Serotonin does not surge in dramatic spikes. It modulates tone. It shapes **pacing**. It smooths extremes and helps emotional waves return toward center.

Where dopamine encodes novelty, serotonin encodes continuity.

Its influence is slower and more diffuse, adjusting the sensitivity of neural circuits over time. It contributes to the regulation of mood amplitude and the modulation of stress responses. It affects sleep timing, appetite, impulse control, and emotional resilience.

Serotonin operates less like a flash of light and more like a dimmer switch.

Its oscillatory behavior is subtle but crucial. When serotonin tone is well regulated, emotional fluctuations remain within adaptive range. When its modulation weakens, amplitude can flatten into apathy or swing into volatility.

Serotonin does not simply "increase happiness."

It stabilizes rhythm.

3.3 Cortisol Waves

Cortisol is often framed as the stress hormone.

But cortisol follows one of the most elegant daily oscillations in the body.

It rises before waking — anticipating the need for alertness. It peaks in the early morning. Then it gradually declines across the day, reaching its lowest point at night.

This is not a stress response. It is a circadian rhythm.

Cortisol helps regulate energy availability, immune activity, and cognitive readiness. Its daily wave prepares the body for engagement and permits recovery in darkness.

Acute stress produces additional pulses — short-term elevations that mobilize resources. In healthy systems, these pulses resolve quickly, allowing the underlying daily wave to remain intact.

When recovery fails or chronic stress persists, the cortisol curve flattens or shifts. The morning peak blunts. Evening levels rise. Sleep fragments. Immune regulation falters.

Cortisol is not inherently harmful.

It is a timed oscillator whose health depends on amplitude, phase, and recovery.

3.4 Melatonin Phase

If cortisol signals day, melatonin signals night.

Melatonin does not induce sleep by force. It marks **phase**. It announces that darkness has arrived and that physiological systems should transition toward rest, repair, and metabolic recalibration.

Its release is tightly coupled to light exposure. In darkness, melatonin rises. In light, it falls.

This oscillation provides temporal structure. It synchronizes sleep cycles, temperature regulation, and hormonal timing.

When artificial light extends into the night, melatonin release delays. Internal rhythms drift. Sleep onset shifts. Downstream systems — immune activity, glucose metabolism, cognitive restoration — follow the misalignment.

Melatonin's power lies not in its concentration, but in its timing.

It is a phase oscillator — a signal that coordinates the transition between day and night across the organism.

3.5 Clock Genes and Internal Time

Beneath hormones and neurotransmitters lies a deeper layer of rhythm.

Inside nearly every cell are molecular clocks — networks of genes that turn on and off in approximately 24-hour cycles. These **clock genes** regulate the timing of protein production, metabolism, cell repair, and immune function.

They form feedback loops: certain genes activate others, which then inhibit the first group, creating a self-sustaining oscillation.

This genetic cycling generates internal time.

The master clock in the brain's suprachiasmatic nucleus synchronizes peripheral clocks throughout the body. Light exposure calibrates the master clock. Food timing and activity influence peripheral clocks. Together they maintain systemic coherence.

If these genetic oscillators fall out of alignment — through shift work, chronic stress, irregular sleep — coordination weakens. Metabolic processes mistime. Immune function desynchronizes. Mood regulation shifts.

Clock genes reveal something profound:

Rhythm is not imposed from outside.
It is encoded at the molecular level.

From Molecules to Motion

Dopamine pulses.
Serotonin modulates.
Cortisol waves.
Melatonin shifts phase.
Clock genes oscillate within cells.

These are not static substances.

They are timed signals.

They carry information through **rise and fall**, not through constant presence. Their meaning depends on amplitude, frequency, phase, synchronization, recovery, and baseline.

When we begin to see neurotransmitters and hormones as oscillators rather than chemicals, the body changes from a container of substances into a network of moving patterns.

And once we see this clearly, a unifying insight emerges:

Life is not regulated by levels alone.

It is governed by rhythm.

Chapter 4 — The Body as a Rhythmic Network

When Waves Begin to Speak to One Another

If each system in the body oscillates, the deeper question is this:

Do they oscillate alone?

It is tempting to imagine that the heart keeps its rhythm independently, that hormones pulse on their own schedule, that immune cells rise and fall in isolation. But biology is not a collection of solo performances. It is a network of interacting waves.

Sleep alters immunity.
Stress reshapes metabolism.
Light influences mood.
Social contact changes heart rate.

These are not secondary effects. They are evidence of coordination.

Every oscillator in the body communicates.

The circadian clock in the brain synchronizes peripheral clocks in the liver, pancreas, and immune cells. The autonomic nervous system links breath to heartbeat. Hormones released in milliseconds influence gene expression that unfolds across hours. Neural rhythms entrain to the voices and facial expressions of others.

The body behaves less like a machine with separate components and more like a field of coupled oscillators — waves influencing waves, rhythms aligning with rhythms.

In physics, when independent pendulums share a surface, they gradually synchronize. In music, when instruments listen to one another, harmony emerges. In biology, synchronization is not decorative — it is essential. Without coordination, energy is wasted. Signals collide. Recovery falters.

Health is not the perfection of any single rhythm.

It is the coherence of many.

This chapter explores that coherence.

We will see how sleep recalibrates immune cycling.
How metabolism anticipates feeding through internal clocks.
How stress follows oscillatory patterns rather than linear escalation.
How human nervous systems entrain to one another in real time.

And we will begin to understand that when we feel "out of sync," it is often not one system failing, but multiple rhythms drifting apart.

The body is not organized by compartments.

It is organized by timing.

Every system listens to another. Every wave modifies the next. The organism survives because its oscillators are coupled, nested, and adaptive.

By the end of this chapter, a larger picture will come into focus:

The body is not a machine.

It is a coordinated orchestra.

4.1 Sleep–Immune Coupling

Sleep is not simply rest for the brain.

It is recalibration for the entire organism.

When we sleep, neural oscillations slow into coordinated waves — especially during deep, slow-wave sleep. These slow oscillations do more than restore cognition. They synchronize immune signaling, hormonal release, and cellular repair.

During the night, immune cells shift behavior. Certain cytokines — chemical messengers that regulate inflammation — rise in tightly controlled patterns. These signals do not indicate illness; they participate in immune memory consolidation. Just as the brain consolidates learning during sleep, the immune system refines its recognition patterns.

Antibody responses improve after sufficient sleep. Vaccination effectiveness depends partly on sleep quality. Recovery from infection accelerates when sleep cycles remain intact.

Sleep provides temporal structure to immunity.

Cortisol, which suppresses inflammation, falls at night. Melatonin rises. Growth hormone pulses. These coordinated changes create a window in which repair, immune training, and cellular maintenance can proceed efficiently.

When sleep becomes fragmented or misaligned, this coordination weakens.

Cortisol may remain elevated into the night. Inflammatory signals may linger into the day. The immune baseline can shift subtly upward. Over time, what begins as mild sleep disruption may contribute to persistent low-grade inflammation.

The connection is not mystical. It is rhythmic.

Sleep and immunity are coupled oscillators. When one drifts, the other follows.

Oscillatory Mapping — Immune Distortion

Variable	What Changes	What It Feels Like
Amplitude	Weak immune response	Frequent infections
Amplitude (excess)	Exaggerated inflammatory surge	Autoimmune flares
Frequency	Persistent activation cycles	Chronic low-grade inflammation
Recovery	Delayed resolution	Lingering illness
Baseline	Elevated inflammatory tone	Constant "low-level sickness"

Variable	What Changes	What It Feels Like
Synchronization	Misalignment with sleep rhythm	Worsened symptoms after poor sleep

Inflammation is not simply "high" or "low."
It is rhythmic — and its health depends on proper oscillation.

Health depends not only on how long we sleep, but on how well sleep aligns with immune timing.

4.2 Metabolic Timing

Metabolism is often described in terms of calories, nutrients, and enzymes.

But beneath these biochemical pathways lies a clock.

Glucose tolerance varies across the day. Insulin sensitivity peaks during daylight hours and declines at night. The liver's capacity to process nutrients follows circadian rhythms. Even the microbiome shifts composition in daily cycles.

Eating at noon is not metabolically identical to eating at midnight.

The body anticipates feeding.

Peripheral clocks in the liver, pancreas, adipose tissue, and gut coordinate their activity with signals from the central circadian clock in the brain. Light exposure calibrates the master clock. Meal timing influences peripheral clocks. When feeding occurs at expected times, metabolic efficiency improves.

When feeding becomes irregular — late-night eating, shift work, chronic circadian disruption — synchronization falters.

Insulin pulses may misalign with glucose peaks. Fat storage signals may remain elevated. Energy oscillations flatten or spike unpredictably.

Over time, desynchronization can contribute to metabolic drift — subtle changes in baseline glucose, lipid regulation, and inflammatory tone.

Metabolism is not a static furnace burning fuel at a constant rate.

It is a timed rhythm responding to environmental cues.

The health of metabolism depends on **phase alignment**, **frequency stability**, and **synchronization** with sleep and light cycles.

When timing is preserved, energy flows predictably.
When timing drifts, fatigue, weight fluctuation, and metabolic instability can emerge.

Sleep and metabolism do not function independently.

They are nested rhythms within the same network.

And the more closely we examine them, the clearer the pattern becomes:

The body organizes energy through time.

Not just through chemistry — but through rhythm.

Oscillatory Mapping — Metabolic Distortion

Variable	What Changes	What It Feels Like
Phase	Eating misaligned with circadian rhythm	Afternoon crashes
Amplitude	Blunted insulin response	Energy instability
Frequency	Erratic glucose cycling	Sugar cravings
Recovery	Slow return to metabolic baseline	Post-meal fatigue
Baseline	Elevated fasting glucose	Persistent energy dullness
Synchronization	Desynchronization with sleep timing	Weight drift

Metabolism does not fail suddenly. It drifts when rhythm drifts.

4.3 Stress Oscillation

Stress is often described as something we either "have" or "don't have."

But stress is not a static condition.
It is a **cycle**.

A healthy stress response follows a recognizable pattern: **activation**, **peak**, and **resolution**.

When a challenge appears, the **sympathetic nervous system** activates. Heart rate rises. Attention sharpens. **Cortisol** and **adrenaline** pulse into circulation. Energy mobilizes. The system prepares for action.

Then, once the challenge passes, the system should shift.

The **parasympathetic nervous system** increases activity. Heart rate slows. Cortisol declines. Muscles relax. The body returns toward **baseline**.

This rise-and-fall pattern is **stress oscillation**.

The key variable is not whether stress occurs.
The key variable is **recovery**.

When stress oscillates cleanly — strong activation followed by efficient return — resilience builds. The system remains flexible. Each cycle strengthens adaptive range.

When activation persists without resolution, oscillation narrows. **Recovery shortens. Baseline drifts upward**. The body begins the next stress cycle already elevated.

Over time, what was once a wave becomes a plateau.

Chronic tension.
Persistent vigilance.
Fatigue layered on top of arousal.

The problem is not stress itself.
It is distorted oscillation — impaired **amplitude control**, accelerated **frequency**, incomplete **recovery**, and elevated **baseline**.

Stress is healthy when it moves.
It becomes harmful when it freezes.

4.4 Social Entrainment

Human beings do not regulate alone.

When two people sit across from each other and speak calmly, their breathing patterns subtly align. Their heart rates begin to synchronize. Neural activity shifts in parallel patterns. This phenomenon is called **entrainment** — the tendency of oscillators to align when connected.

In social interaction, this becomes **social entrainment**.

The **vagus nerve** transmits signals between facial expression, vocal tone, heart rhythm, and emotional state. When someone speaks gently, their vocal cadence influences another person's autonomic system. Calm can spread. So can agitation.

This is not metaphor. It is physiology.

Infants regulate their nervous systems through caregivers' rhythms. Adults regulate through eye contact, touch, shared silence, and synchronized movement. Even group experiences — music, collective rituals, coordinated activity — amplify entrainment across many individuals.

Social connection modulates **amplitude** of emotion, stabilizes **frequency** of stress cycles, and improves **recovery** efficiency. It lowers cortisol and influences immune tone.

Isolation disrupts this regulation.

Without co-regulation, the nervous system must bear its own fluctuations alone. **Baseline tension** may rise. **Stress frequency** may accelerate. Sleep and metabolic rhythms may desynchronize.

Belonging is not merely psychological comfort.

It is rhythmic stabilization.

4.5 Why Systems Speak to Each Other

Why does sleep influence immunity?
Why does stress affect metabolism?
Why does loneliness alter inflammation?

Because oscillators do not exist in isolation.

In complex systems, coupling increases efficiency. When rhythms align, energy transfer improves. Signals reinforce rather than collide.

The body's systems share information through:

- **Hormonal signaling**
- **Neural feedback loops**
- **Immune messengers**
- **Circadian timing signals**

When one oscillator shifts, others detect the change and adjust.

If sleep shortens, cortisol timing shifts.
If cortisol shifts, immune baseline changes.
If immune tone rises, metabolic regulation adapts.
If metabolism drifts, neural energy availability alters mood.

This is not fragility.

It is interdependence.

Communication between systems allows coordinated adaptation. But it also means that persistent distortion in one rhythm can gradually spread.

The body behaves as a **network of coupled oscillators** — nested, synchronized, adaptive.

Health emerges from coordinated fluctuation across systems.

Illness often reflects breakdown in coordination.

End of Part I. Health Is Structured Oscillation, Not Static Stability

By now, a new image of the body should be clear.

The heart varies.
Hormones pulse.
Neurons oscillate.
Immunity cycles.
Stress rises and resolves.
Metabolism follows time.
Social contact synchronizes physiology.

Each system is rhythmic.

Each rhythm influences the others.

The body is not a machine composed of independent parts.

It is a field of interacting waves.

It is not engineered like gears.

It performs like music.

The body is not a machine.

It is a **coordinated orchestra**.

Glossary — Part I: The Oscillating Organism

Amplitude

The magnitude of rise and fall within a biological rhythm.
High amplitude reflects strong activation followed by clear descent.
Low amplitude reflects flattened or blunted response.
Excess amplitude reflects exaggerated peaks and troughs.

Baseline

The resting set-point around which oscillations occur.
A stable baseline allows adaptive fluctuation.
A shifted baseline (too high or too low) alters the starting point of every subsequent cycle.

Circadian Rhythm

A roughly 24-hour biological oscillation synchronized primarily by light exposure.
Regulates sleep–wake timing, cortisol release, metabolism, immune function, and cellular repair.

Coupled Oscillators

Biological rhythms that influence one another through shared signals.
Examples include sleep–immune coupling, heart–breath synchronization, and hormonal–neural coordination.

Dynamic Regulation

The principle that biological stability is maintained through controlled fluctuation rather than rigid constancy.
True stability involves continuous micro-adjustment.

Entrainment

The synchronization of one rhythm to another external or internal rhythm.
Examples include circadian alignment with light and social co-regulation of nervous systems.

Frequency

The rate at which a biological cycle repeats over time.
Healthy frequency allows sufficient activation and sufficient recovery.
Distorted frequency may be excessively rapid (anxiety) or excessively slowed (lethargy).

Heart Rate Variability (HRV)

The millisecond variation between heartbeats.
Higher variability reflects adaptive autonomic flexibility.
Reduced variability suggests rigidity and stress burden.

Homeostasis
Traditionally defined as internal stability.
In living systems, homeostasis is achieved through dynamic oscillation rather than static constancy.

Immune Cycling
The patterned rise and fall of immune activity, including circadian variation in cytokines and controlled inflammatory responses.

Neural Oscillations
Rhythmic electrical activity in the brain (e.g., alpha, beta, delta waves) that coordinate cognition, perception, and emotional processing.

Oscillation
A repeating rise-and-fall movement around a baseline.
The foundational architecture of living systems.

Oscillatory Health Model
A framework describing biological rhythms using six variables:
Amplitude, Frequency, Phase, Synchronization, Recovery, and Baseline.
Health reflects flexible modulation of these variables.

Phase
The timing of a rhythm relative to environmental cues (light, food, activity).
Proper phase alignment ensures energy and recovery occur at appropriate times.

Recovery
The efficiency and completeness with which a system returns to baseline after activation.
Impaired recovery leads to chronic elevation or incomplete resolution.

Synchronization
The coordination between multiple biological rhythms.
Health depends on systems rising and falling in temporal harmony.

Stress Oscillation

The natural cycle of activation and resolution in response to challenge. Healthy stress includes a clean rise and a complete return.

Rhythmic Network
The concept that the body functions as an interconnected field of oscillators rather than isolated systems.

Variability
Adaptive fluctuation within safe physiological ranges.
Reduced variability often precedes dysfunction.

Wave
A visual metaphor for biological oscillation — rise, peak, fall, and return.

PART II — Rhythm as Medicine

When the Wave Heals

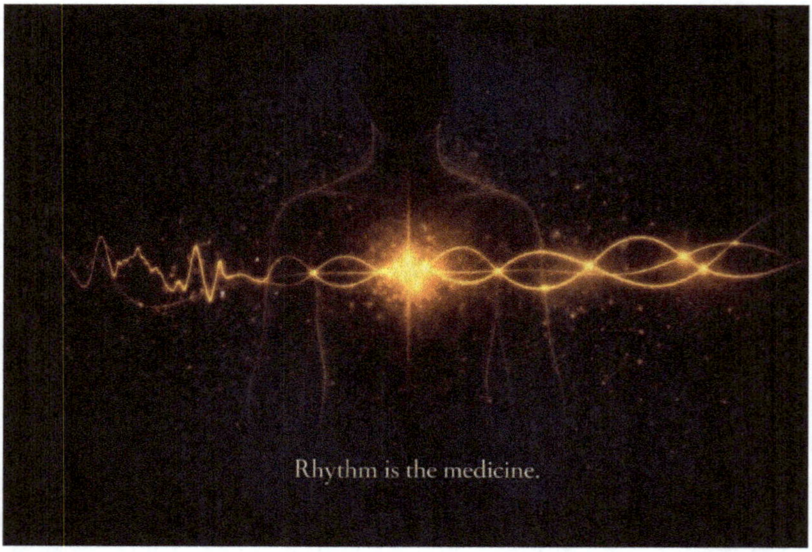

If Part I revealed the architecture of oscillation, Part II explores its restoration.

Until now, we have examined structure — amplitude, frequency, phase, synchronization, recovery, baseline. We have seen how distortions in these variables can quietly erode vitality. But the deeper truth is this:

Rhythm does not only explain illness.
Rhythm is often the medicine.

Modern medicine frequently approaches disease as something to suppress — lower the inflammation, block the signal, silence the symptom, stabilize the level. And sometimes that is necessary. But beneath suppression lies a subtler and more powerful approach: restore the wave.

Because health rarely requires elimination of oscillation.
It requires recalibration.

When sleep realigns with light, immune tone recalibrates.
When emotional amplitude regains range, resilience returns.

When metabolic timing synchronizes with circadian rhythm, energy stabilizes.
When social connection restores entrainment, stress recovers more efficiently.

In each case, the intervention is not mechanical correction.
It is rhythmic restoration.

This section of the book will move from abstraction to lived experience.

Each chapter begins with a human story — not because narrative replaces science, but because rhythm is felt before it is measured. We will observe how distortion appears in everyday life: insomnia, emotional flattening, lingering inflammation, metabolic fatigue, social isolation.

Then we will identify the **primary variable** involved.
Is it **phase misalignment**?
Is it impaired **recovery**?
Is it narrowed **amplitude**?
Is it elevated **baseline**?
Is it fractured **synchronization**?

And finally, we will explore how restoration occurs — not through force, but through realignment of rhythm.

Sleep becomes a master oscillator.
Emotion becomes a wave to be ridden, not suppressed.
Inflammation becomes a pulse that must resolve.
Metabolism becomes a timed choreography.
Belonging becomes physiological entrainment.

The goal of this part is not to offer quick fixes.

It is to show what healthy oscillation actually looks like.

To recognize that vitality is dynamic.
That sadness can be adaptive.
That stress can strengthen when recovery is intact.
That fever can heal.
That timing can restore.

If Part I was about structure, Part II is about coherence.

And coherence is medicine.

Because when rhythms are allowed to move — to rise, fall, align, and recover — the organism often remembers how to heal itself.

By the end of this section, one insight should become clear:

Medicine is not always about suppressing waves.

It is often about restoring them.

Chapter 5 — Sleep as Master Oscillator

Where the Body Resets Itself

He thought the problem was fatigue.

But fatigue was only the surface.

The deeper issue was timing.

For months, he had gone to bed at irregular hours — sometimes midnight, sometimes two in the morning, occasionally earlier when exhaustion forced it. Screens filled the last hour before sleep. Light lingered long after sunset. Weekends drifted later still.

He still slept six or seven hours most nights.

But something was off.

Morning felt heavier. Minor infections lingered. Small stressors felt amplified. Recovery from exercise slowed. Mood dulled more quickly after effort.

Nothing dramatic. Nothing diagnosable.

Just subtle erosion.

The rhythm had shifted.

Sleep is often treated as a passive state — the absence of waking. But biologically, sleep is one of the most active periods in the human organism. It is not a shutdown. It is recalibration.

During sleep, neural oscillations reorganize memory.
Hormonal pulses shift dramatically.
Immune signaling recalibrates.
Metabolic processes reset.
Cortisol falls.
Melatonin rises.
Growth hormone surges.

Sleep is not just rest.

It is the master oscillator.

Among all the rhythms in the body, sleep holds a unique coordinating role. It anchors **phase** — aligning internal clocks with the external light–dark cycle. It protects **recovery** — ensuring that activation cycles resolve fully before the next day begins.

When sleep timing aligns and recovery remains intact, downstream systems function coherently. Immunity resolves inflammation efficiently. Metabolism anticipates feeding. Mood maintains range. Stress cycles complete.

When sleep drifts — even slightly — these systems begin to desynchronize.

The distortion may begin with **phase**: going to bed too late relative to internal circadian timing. Or with impaired **recovery**: fragmented sleep that prevents deep oscillatory restoration.

At first, nothing collapses.

But gradually, the body begins to accumulate incomplete cycles.

Stress does not fully resolve.
Immune activation lingers.
Cortisol curves flatten.
Emotional amplitude narrows.

Sleep is not merely one rhythm among many.

It is the conductor.

In this chapter, we will explore why deep sleep architecture matters, how nighttime immune recalibration occurs, how the brain clears metabolic waste during slow-wave cycles, how light anchors phase alignment, and what happens when recovery repeatedly fails.

To understand sleep is to understand rhythm itself.

Because sleep is where the body practices return — return to baseline, return to coherence, return to timing.

And without return, oscillation becomes distortion.

Sleep is not the absence of activity.

It is the foundation of restoration.

5.1 The Architecture of Deep Sleep

Not all sleep is the same.

The body does not simply turn off and remain in one static state. Instead, it moves through **stages**, each with distinct oscillatory patterns. Among these, **deep sleep** — also called **slow-wave sleep** — is structurally unique.

During deep sleep, the brain produces large, synchronized **delta waves**. These slow oscillations sweep across the cortex in coordinated pulses. Neurons fire together, then fall silent together. Rise. Fall. Rise. Fall.

This synchronized activity is not random. It creates windows for cellular repair, memory consolidation, hormonal release, and metabolic recalibration.

At the same time, **growth hormone** is released in pulses, supporting tissue repair and immune modulation. **Cortisol** remains low, allowing inflammatory signals to quiet. Heart rate slows. Blood pressure declines. Energy demand drops.

Deep sleep is the body's most powerful expression of **recovery**.

Its architecture depends on amplitude and synchronization — large coordinated neural waves that create physiological stillness without rigidity. When deep sleep shortens or fragments, recovery narrows. The following day begins from a partially restored baseline.

The body does not need perfect sleep.

But it does require sufficient deep oscillation to complete its nightly reset.

5.2 Immune Recalibration at Night

While consciousness rests, the immune system becomes active — not in alarm, but in regulation.

Immune cells follow circadian patterns. Certain cytokines rise during sleep, supporting immune memory and adaptive recalibration. The timing of these signals is critical. They are meant to increase at night and decline toward morning.

Sleep provides the environment for this recalibration.

With **cortisol** low and **parasympathetic tone** elevated, inflammatory activity can rise temporarily without escalating into chronic activation. The system practices controlled oscillation — activation followed by resolution.

Studies show that vaccination response improves after adequate sleep. Infection recovery accelerates when sleep cycles remain intact. Even minor sleep restriction shifts inflammatory markers upward.

The immune system does not simply defend.

It cycles.

Nighttime is when that cycling refines itself.

When sleep becomes irregular or truncated, immune **baseline** may drift upward. Low-grade inflammation persists into daytime. Recovery shortens. The next immune challenge begins from an already elevated state.

The distortion is subtle — but cumulative.

Immune health depends not only on strength of response, but on **timed oscillation**.

5.3 Glymphatic Clearance

During wakefulness, the brain consumes enormous energy. Neurons fire rapidly. Metabolic byproducts accumulate. Unlike other organs, the brain lacks traditional lymphatic vessels.

So how does it clear waste?

Primarily during sleep.

In deep sleep, the spaces between brain cells expand slightly. Cerebrospinal fluid flows more freely through neural tissue in rhythmic pulses. This process, known as the **glymphatic system**, clears metabolic byproducts — including proteins that accumulate during neural activity.

This clearance depends on oscillation.

Slow-wave activity enhances fluid movement. Heartbeat and respiration subtly modulate cerebrospinal flow. The brain's cleaning mechanism is synchronized with sleep stage transitions.

When deep sleep diminishes, clearance efficiency declines. Waste removal slows. Over time, incomplete nightly reset may contribute to cognitive fog, reduced mental resilience, and possibly long-term neurodegenerative risk.

The glymphatic system is not active continuously.

It depends on phase-appropriate, amplitude-rich deep sleep.

Sleep is not passive unconsciousness.

It is structured, rhythmic maintenance.

In deep oscillation, the brain repairs itself.
In nightly immune recalibration, inflammation resets.
In glymphatic clearance, metabolic waste is removed.

Together, these processes reveal something essential:

Sleep is not optional downtime.

It is the biological expression of **recovery** — the nightly restoration of oscillatory integrity.

5.4 Light as Phase Anchor

If sleep is the master oscillator, **light** is its anchor.

The body does not keep time in isolation. It calibrates itself daily against the most powerful environmental cue available: the light–dark cycle.

In the brain, a small structure called the **suprachiasmatic nucleus (SCN)** acts as the master clock. It receives direct input from the eyes — not visual information in the ordinary sense, but data about brightness and timing. Morning light tells the clock: the day has begun. Evening darkness signals: it is time to prepare for rest.

This daily exposure to light aligns **phase**.

When light arrives at the appropriate time — early in the day — cortisol rises sharply, alertness increases, and internal clocks synchronize. When darkness deepens at night, **melatonin** rises, lowering body temperature and shifting physiology toward sleep and repair.

But modern environments blur these signals.

Artificial light extends the day. Screens emit blue wavelengths that suppress melatonin. Late-night brightness delays internal timing. The body receives contradictory messages: biological night has arrived, yet light suggests day.

When **phase** drifts, sleep onset shifts later. Cortisol timing misaligns. Immune rhythms lose coherence. Metabolic clocks desynchronize from feeding behavior.

Light is not merely illumination.

It is instruction.

Without consistent light cues, internal clocks free-run, drifting gradually out of alignment. Small phase delays accumulate into persistent circadian disruption.

Anchoring phase does not require perfection. It requires consistency.

Morning light exposure.
Darkness at night.
Regular sleep timing.

These are not lifestyle preferences. They are phase regulators — daily recalibrations that maintain rhythmic coherence across the organism.

Light keeps the orchestra in time.

5.5 When Recovery Fails

The most important question in sleep is not only how long we sleep.

It is whether we **recover**.

A healthy system activates during the day and resolves at night. Cortisol rises, then falls. Immune activation peaks when needed and quiets afterward. Neural firing intensifies during wakefulness and reorganizes during sleep.

When recovery fails, the next cycle begins from an incomplete reset.

Fragmented sleep shortens deep slow-wave periods. Cortisol may remain elevated into the evening. Inflammatory signals may persist. Heart rate variability narrows. Emotional amplitude dampens.

The system never fully returns to **baseline**.

At first, the body compensates.
But compensation narrows flexibility.

Days of partial recovery become weeks. Weeks become months. Baseline shifts slightly upward — tension persists, energy remains dull, mood fluctuates more quickly.

What began as minor sleep disruption becomes systemic drift.

Impaired recovery alters multiple variables at once:

- **Amplitude** narrows — energy dulls.
- **Frequency** accelerates — stress cycles shorten.
- **Phase** misaligns — timing drifts.
- **Synchronization** weakens — systems lose coherence.
- **Baseline** rises — activation never fully resolves.

Recovery is the hinge upon which oscillatory health turns.

Without return, rhythm becomes distortion.
Without restoration, oscillation becomes rigidity.

Sleep protects recovery.

When sleep stabilizes — in timing, depth, and continuity — the body regains flexibility. Cortisol curves normalize. Immune pulses recalibrate. Emotional range widens. Metabolic timing improves.

Oscillatory Mapping — Sleep Distortion

Variable	What Changes	What It Feels Like
Amplitude	Reduced deep slow-wave intensity	Unrestorative sleep
Frequency	Fragmented cycles	Frequent awakenings
Phase	Delayed circadian timing	Difficulty falling asleep
Synchronization	Misalignment with light–dark cycle	Jet-lagged feeling
Recovery	Incomplete nightly reset	Morning fatigue

Variable	What Changes	What It Feels Like
Baseline	Elevated nighttime cortisol	Light, restless sleep

Sleep disruption is rarely a single problem. It is often a multi-variable distortion.

The restoration of sleep is often the restoration of rhythm itself.
Sleep does not merely give energy.
It protects oscillation.
And when oscillation is preserved, the organism remembers how to heal.

Chapter 6 — Emotional Oscillation and Resilience

The Strength of the Wave

She thought something was wrong with her because she felt too much.

Joy lifted her quickly. Disappointment cut sharply. After achievement came a strange quiet. After connection came a soft ache. She watched others who seemed steady — unshaken — and wondered why her emotional life moved in such visible arcs.

But what if the problem was not that she felt too deeply?

What if the problem was that we have been taught to fear amplitude?

Emotion, like every other biological system, is rhythmic.

It rises.
It peaks.
It falls.
It returns.

Yet culturally, we are conditioned to equate health with emotional flatness — steady mood, controlled reaction, minimized fluctuation. We are told to avoid the high, suppress the low, smooth the curve.

But the nervous system does not define resilience as emotional sameness.

It defines resilience as **adaptive amplitude**.

In *The Happiness Rebound*, we explored how joy itself carries a built-in descent — how dopamine surges must eventually resolve, how the brain recalibrates after pleasure, how meaning emerges from the return. That book traced the rise and fall of happiness.

This chapter expands that insight.

Emotion is not meant to remain elevated. Nor is it meant to remain suppressed. It is designed to move.

Healthy sadness is not pathology. It is integration.
Healthy joy is not excess. It is expansion.
Healthy stress is not damage. It is mobilization followed by rebound.

The key variable here is **amplitude** — the strength of the rise and fall.

When amplitude narrows, life feels muted.
When amplitude becomes uncontrolled, volatility replaces coherence.
When amplitude remains flexible, resilience emerges.

Emotional health does not require constant positivity.
It requires oscillation with recovery intact.

In this chapter, we will examine how sadness can be adaptive, how dopamine pulses shape emotional arcs, how stress rebounds build capacity, and how psychological flexibility depends on preserving range rather than eliminating fluctuation.

Resilience is not the absence of emotion.

It is the ability to move through emotion and return.

And sometimes, what we mistake for instability is simply the nervous system doing what it was designed to do — rising, falling, and remembering how to come back.

Emotional strength is not flatness.

It is a wave that knows how to return to shore.

6.1 Healthy Sadness

Sadness has been misunderstood.

We treat it as a symptom to eliminate, a deviation from health, a sign that something inside us is malfunctioning. But sadness, in its natural form, is not a failure of the emotional system.

It is a phase of oscillation.

When something meaningful is lost — a relationship, an expectation, a chapter of life — the nervous system shifts. Energy decreases. Attention turns inward. Social drive softens. The body slows slightly. This is not collapse. It is recalibration.

In oscillatory terms, sadness represents the downward arc of emotional amplitude.

Just as joy cannot remain at peak indefinitely, neither can sadness persist at depth without consequence. The key variable is not the presence of sadness, but its **range** and its **recovery**.

Healthy sadness has structure:

- It arises in response to meaningful change.
- It reduces outward drive temporarily.
- It encourages reflection and integration.
- It gradually resolves.

This temporary lowering of emotional amplitude allows the nervous system to conserve energy, reassess priorities, and encode experience into memory. During this period, serotonin modulation supports stabilization, while stress hormones gradually decline.

Sadness is biologically adaptive because it slows the system enough to integrate what has happened.

When sadness is suppressed prematurely, the oscillation becomes incomplete. When it is prolonged without recovery, amplitude narrows and baseline shifts downward.

Resilience does not require avoiding the downward arc.

It requires trusting that the arc will return.

Healthy sadness is not the opposite of happiness.

It is part of the same wave.

6.2 Dopamine Rise and Fall

If sadness represents the descent of amplitude, dopamine represents the ascent.

Dopamine pulses when possibility appears — when anticipation rises, when effort seems promising, when reward feels near. It creates upward momentum. It sharpens attention and energizes action.

But dopamine does not sustain elevation indefinitely.

After achievement, the pulse declines. After anticipation resolves, receptors recalibrate. The nervous system reduces sensitivity to prevent overstimulation.

This is not punishment.

It is rhythm.

In *The Happiness Rebound*, we explored how every emotional high carries a built-in recalibration. The brain must descend from peak activation to preserve sensitivity. Without this fall, motivation would lose contrast. Pleasure would flatten into noise.

The rise-and-fall pattern of dopamine defines emotional amplitude.

A healthy dopamine system:

- Rises clearly in anticipation.
- Peaks in engagement.
- Falls after resolution.
- Allows recovery before the next rise.

When this oscillation is intact, life feels dynamic but not chaotic.

When amplitude narrows, motivation dulls.
When amplitude spikes excessively, volatility increases.
When recovery shortens, the next rise begins from depletion.

The mistake is not in experiencing the fall.

The mistake is interpreting the fall as failure.

The descent after joy is not a flaw in the system.

It is the system preserving flexibility.

Emotional resilience depends on allowing dopamine to rise and fall cleanly — not chasing perpetual elevation, not suppressing descent, but trusting the oscillation.

Because without the fall, the next rise cannot occur.

And without amplitude, emotion loses depth.

Resilience is not constant elevation.

It is intact oscillation.

6.3 Stress and Rebound

Stress is often portrayed as the enemy of emotional health.

But stress, in its natural form, is a wave.

A challenge appears. The nervous system activates. Heart rate rises. Cortisol pulses. Attention sharpens. Energy mobilizes. This upward surge is not damage — it is preparation.

Then comes the crucial moment.

Does the system return?

In healthy oscillation, stress is followed by **rebound**.

Cortisol declines.
Parasympathetic tone increases.
Breathing slows.
Emotional intensity softens.

The system settles.

This rebound phase is where resilience is built.

Without activation, growth does not occur. Without recovery, depletion accumulates.

In oscillatory terms, stress requires intact **amplitude** and intact **recovery**. The rise must be strong enough to mobilize. The fall must be complete enough to restore baseline.

When rebound fails — when activation lingers — emotional range narrows. Baseline tension rises. The next stressor begins from partial elevation. Over time, the nervous system forgets what full descent feels like.

Resilience is not the absence of stress.

It is the preservation of rebound.

Just as we explored in *The Happiness Rebound*, emotional peaks require descent. The same is true for stress activation. The fall is not weakness. It is recalibration.

The wave strengthens when both directions are allowed.

6.4 Psychological Flexibility

If amplitude describes the strength of the wave, psychological flexibility describes the mind's ability to move with it.

Flexibility is the capacity to experience emotion without becoming trapped in it. It is the ability to expand during joy and contract during sadness without losing coherence. It is the willingness to feel discomfort while still acting according to values.

Rigid systems resist oscillation.

They attempt to eliminate sadness, suppress anxiety, prolong pleasure, avoid discomfort. But suppression flattens amplitude. Avoidance narrows range. Over-control reduces adaptability.

Flexible systems, by contrast, allow oscillation while preserving structure.

They do not panic at emotional descent.
They do not chase perpetual elevation.
They trust that waves resolve.

Psychological flexibility is the cognitive counterpart of physiological variability. Just as heart rate variability signals autonomic adaptability, emotional variability signals resilience — when recovery remains intact.

Flexibility does not mean instability.

It means controlled range.

In rhythmic terms, flexibility preserves **amplitude** without losing **recovery**.

6.5 Emotional Range as Health

We often equate happiness with emotional health.

But health is not constant positivity.

It is **range**.

Range allows joy to feel vivid.
Range allows sadness to feel meaningful.
Range allows stress to mobilize and resolve.
Range allows calm to be restorative.

In *The Happiness Rebound*, we examined how the pursuit of sustained elevation often backfires. Dopamine peaks require recalibration. The fall is not a flaw — it is structural necessity. Meaning emerges in the descent, not in perpetual ascent.

This chapter extends that insight:

Emotional health is not the elimination of negative states.
It is the preservation of oscillation.

When amplitude flattens, life feels muted.
When amplitude becomes chaotic, stability dissolves.
When amplitude remains flexible — rising, falling, returning — resilience emerges.

Emotional range is not volatility.

It is vitality.

Oscillatory Mapping — Emotional Distortion

Variable	What Changes	What It Feels Like
Amplitude	Flattened affect	Emotional numbness
Amplitude (excess)	Exaggerated swings	Mood volatility
Frequency	Rapid cycling	Anxiety, agitation
Recovery	Prolonged emotional activation	Rumination
Baseline	Downward shift	Persistent low mood
Synchronization	Misalignment with stress or sleep cycles	Emotional burnout

Emotional health is not constant elevation. It is preserved oscillation with intact recovery.

The goal is not to hold the wave at its peak.
The goal is not to eliminate its trough.

The goal is to maintain the integrity of the oscillation.

Because happiness that never descends loses contrast.
Sadness that never resolves becomes pathology.
Stress that never rebounds becomes exhaustion.

Resilience lives in the full arc.

And the arc, as we learned before, is not a defect in the system.

It is the design.

Emotion, like every biological rhythm, is healthiest when it moves — and when it remembers how to return.

Chapter 7 — Immune Pulsatility and Inflammatory Cycles

The Fire That Must Rise and Fall

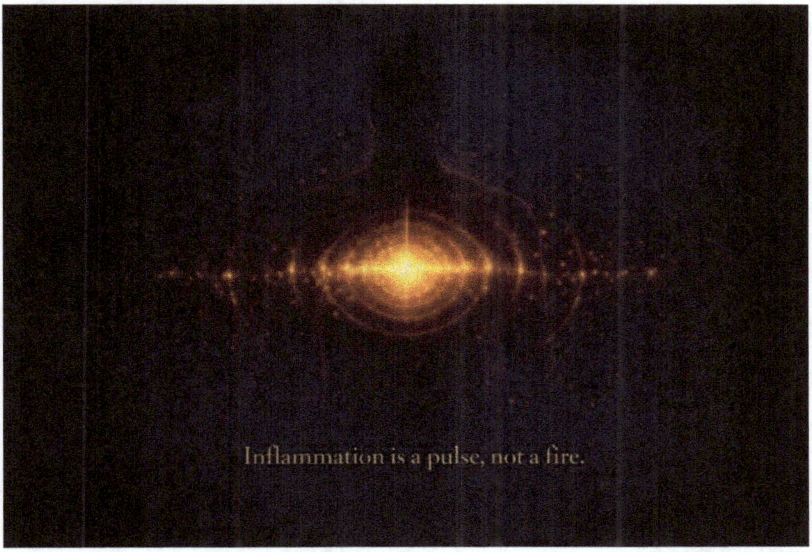

He thought his body was attacking itself.

Joint stiffness in the morning.
Low-grade fatigue.
A persistent sense of "inflammation" that never fully disappeared.

Blood tests showed markers slightly elevated — not alarming, but not quiet. Nothing dramatic enough to define as crisis. Yet something was clearly unsettled.

The common narrative would describe this as an immune problem — too much inflammation, too much activation, too much fire.

But inflammation is not inherently harmful.

Inflammation is a wave.

When the body encounters injury or infection, immune cells activate. **Cytokines** rise. Blood vessels dilate. Heat increases. Energy redirects toward defense and repair. This surge is purposeful. It is necessary. It is intelligent.

Then — in healthy oscillation — it resolves.

Signals shift from activation to containment. Repair processes take over. Cytokine levels decline. Tissue recalibrates. The immune system returns toward **baseline**.

This rise-and-fall pattern is **immune pulsatility**.

The problem is not activation.

The problem is distortion of the variables.

In acute inflammation, **frequency** is appropriate and temporary. The wave rises when needed and quiets afterward.

In chronic inflammation, either the wave repeats too frequently, or the **baseline** shifts upward. The system begins from a state of partial activation. Recovery shortens. The immune system forgets what full quiet feels like.

Inflammation is not an enemy to eliminate.

It is a rhythm to regulate.

This chapter explores how immune activity naturally cycles, how cytokines follow daily rhythms, how fever represents controlled oscillation, and how subtle baseline shifts can quietly reshape physiology across months or years.

We will see that immune health is not defined by the absence of activation.

It is defined by the preservation of pulsatility.

The fire must rise when needed.
It must fall when the work is done.

When it does both, health is maintained.

When it cannot fall, the system begins to burn from within.

Immune resilience is not about extinguishing the flame.

It is about restoring the wave.

7.1 Acute vs Chronic Inflammation

Inflammation has acquired a negative reputation.

We associate it with swelling, pain, autoimmune disease, cardiovascular risk, aging. The word itself sounds pathological.

But inflammation, in its original form, is protective.

Acute inflammation is a tightly regulated oscillation. When tissue is injured or pathogens invade, immune cells detect danger signals and release **cytokines**. Blood flow increases. Immune cells migrate to the site. Repair processes activate.

The response is strong — sometimes visibly dramatic.

Then, just as importantly, it resolves.

Resolution is not passive fading. It is an active phase. Specialized signals turn off pro-inflammatory pathways and promote tissue repair. Immune cells shift phenotype. Debris is cleared. Baseline is restored.

The oscillation completes.

In this pattern, **frequency** is appropriate and episodic. The immune system activates when needed and quiets when finished. The **baseline** remains low between events.

Chronic inflammation, by contrast, represents distortion of this rhythm.

Activation may be less intense than in acute infection, but it fails to resolve completely. Cytokine levels remain slightly elevated. Immune cells maintain a partially activated state. The system begins the next day, or the next challenge, from an elevated **baseline**.

The difference is not simply magnitude.

It is timing.

Acute inflammation is a wave.
Chronic inflammation is a plateau.

In chronic states, either activation cycles repeat too frequently — distorted **frequency** — or the resting set-point has shifted upward — distorted **baseline**.

Over time, this subtle elevation affects metabolism, mood, vascular health, and energy regulation. The immune system, designed for pulsatile defense, becomes persistently vigilant.

The problem is not inflammation itself.

The problem is the loss of oscillation.

7.2 Cytokine Rhythms

Cytokines are often described as inflammatory messengers — chemical signals that amplify immune response.

But they do not simply rise during illness and remain silent otherwise.

They follow rhythms.

Certain cytokines fluctuate across the 24-hour day. Some peak during the night, supporting immune surveillance and tissue repair. Others decline toward morning, coordinating with cortisol's rise. These patterns are synchronized with **circadian timing**.

Cytokines also oscillate in shorter cycles. After activation, pro-inflammatory signals rise quickly. Then anti-inflammatory mediators emerge to contain and resolve the response.

This interplay creates immune pulsatility — alternating waves of activation and regulation.

The significance of this rhythmic behavior is profound.

If cytokines were constant, the immune system would either be perpetually inflamed or perpetually suppressed. Oscillation allows flexibility. It permits strong response when necessary and efficient return afterward.

Distortion occurs when:

- **Frequency** increases — repeated activation without sufficient spacing.
- **Recovery** shortens — incomplete resolution.
- **Baseline** shifts upward — persistent low-grade signaling.

Modern stress, sleep disruption, metabolic imbalance, and environmental factors can subtly alter cytokine timing. The result is not dramatic infection, but chronic low-grade inflammation — a state in which the immune wave never fully descends.

Cytokines are not static markers.

They are rhythmic signals.

Understanding their oscillatory nature changes the question from:

"How do we eliminate inflammation?"

to

"How do we restore immune pulsatility?"

Because immune health depends not on silence, but on timing.

The immune system is healthiest when it knows how to rise — and when it remembers how to fall.

7.3 Fever as Controlled Oscillation

Fever is often feared.

A rising temperature feels alarming — heat, discomfort, weakness. The instinct is immediate suppression. Lower it. Cool it. Stop it.

But fever, in its healthy form, is not a malfunction.

It is a **controlled oscillation**.

When immune cells detect infection, they release cytokines that signal the brain's temperature-regulating centers. The body raises its thermal **set-point** deliberately. Heat increases metabolic activity in immune cells and inhibits replication of certain pathogens. It accelerates defensive processes.

Then, once the threat diminishes, the temperature returns to normal.

Rise.
Peak.
Resolution.

The power of fever lies not in constant elevation, but in its temporality. It is meant to surge and then subside.

In oscillatory terms, fever demonstrates how **baseline** can shift briefly — upward for defense — and then reset. It shows how **amplitude** can increase temporarily in service of recovery.

Problems arise when this oscillation fails to resolve. Persistent low-grade fever, or chronic inflammatory activation without clear cause, reflects distortion of **recovery** or elevation of **baseline**.

The immune system was designed for decisive pulses.

Fever reminds us that activation is not pathology.

Failure to descend is.

7.4 When the Baseline Shifts

Not all immune distortion announces itself dramatically.

Sometimes the most powerful change is subtle.

In healthy oscillation, immune activation begins from a quiet resting tone. After response and resolution, the system returns to that low **baseline**.

But over time — through chronic stress, sleep disruption, metabolic imbalance, environmental exposure — the resting tone can rise slightly.

The shift may be small.

Inflammatory markers remain within "normal" range, yet trend upward. Energy feels marginally lower. Recovery takes slightly longer. Minor stressors provoke disproportionate responses.

The wave still exists.

But it now rises from a higher starting point.

This upward shift in **baseline** is often the earliest stage of chronic inflammatory disease. The immune system becomes persistently vigilant. The body behaves as though mild threat is always present.

Baseline shift alters other systems as well.

Elevated immune tone influences mood regulation, increasing vulnerability to depression. It affects insulin signaling, contributing to metabolic instability. It interacts with stress pathways, narrowing **recovery**.

The distortion is not explosive.

It is cumulative.

The immune system was designed for pulsatility — alternating activation and quiet. When baseline drifts upward and oscillation narrows, flexibility diminishes.

Health is not defined by the absence of immune activity.

It is defined by the preservation of rhythmic return.

Inflammation must rise when necessary.
It must fall when the work is done.
And the resting line must remain adaptive.

When baseline remains flexible, resilience persists.

When baseline shifts upward and fails to reset, chronic illness begins — not as sudden breakdown, but as gradual loss of oscillatory integrity.

Immune health, like emotional health, depends on the wave remembering how to return.

Chapter 8 — Metabolic Timing and Energy Regulation

When Energy Follows the Clock

He believed his problem was willpower.

Afternoon crashes.
Late-night hunger.
Energy spikes after sugar.
Foggy mornings.

He adjusted calories. Changed macros. Tried new supplements. Nothing fully stabilized.

What he never questioned was **timing**.

Metabolism is usually framed as chemistry — carbohydrates, fats, proteins, enzymes, mitochondria. But beneath this biochemical language lies something quieter and more powerful:

Metabolism runs on a clock.

The body does not process energy the same way at midnight as it does at noon. **Insulin sensitivity** fluctuates across the day. **Glucose tolerance** peaks earlier and declines at night. The liver anticipates feeding. Peripheral metabolic tissues coordinate their activity with the brain's master circadian clock.

Energy is not simply consumed.

It is scheduled.

The primary variables that govern metabolic health are **phase** and **synchronization**.

Phase determines whether internal metabolic clocks are aligned with environmental signals — especially light and sleep.
Synchronization determines whether feeding behavior, hormonal pulses, and cellular clocks are coordinated with one another.

When meals align with circadian timing and sleep remains stable, insulin pulses cleanly. Glucose rises and falls predictably. Energy feels steady. Recovery after exertion is efficient.

When timing drifts — irregular meals, late-night eating, inconsistent sleep, shift work — oscillators fall out of alignment. Insulin pulses may arrive at the wrong phase. Glucose variability increases. Fat storage signals persist longer than intended.

The distortion is not always immediate.

But over time, misalignment accumulates.

This chapter explores metabolism not as static fuel management, but as rhythmic choreography. We will examine how **insulin pulses** encode timing, how **meal timing** entrains peripheral clocks, how **glucose variability** reflects oscillatory integrity, how **circadian feeding windows** preserve synchronization, and how desynchronization quietly precedes metabolic disease.

Energy regulation is not only about what enters the body.

It is about when and how oscillators align.

Because metabolism does not simply burn calories.

It moves through time.

And when time is respected, vitality stabilizes.

8.1 Insulin Pulses

Insulin is often described as a storage hormone — a chemical that lowers blood sugar by moving glucose into cells.

But insulin is not a steady drip.

It is released in **pulses**.

Even in the fasting state, the pancreas secretes insulin in small, rhythmic bursts approximately every few minutes. After a meal, these pulses increase in amplitude. Glucose rises. Insulin surges. Cells absorb energy. Then levels fall again.

This pulsatility is essential.

Continuous insulin exposure dulls cellular sensitivity. Pulsed delivery preserves responsiveness. The rise-and-fall pattern allows tissues to distinguish signal from background.

In oscillatory terms, insulin relies on healthy **amplitude**, appropriate **frequency**, and intact **recovery**.

When meals are spaced and aligned with circadian timing, insulin pulses remain clear and distinct. When grazing becomes constant — frequent snacking, late-night eating — pulses blur into prolonged elevation. Recovery shortens. Baseline insulin drifts upward.

Over time, cells respond less efficiently.

The problem is not insulin itself.

It is distortion of its rhythm.

Insulin health depends not only on how much is released, but on whether it rises and falls cleanly.

8.2 Meal Timing

Food is not merely fuel.

It is a timing signal.

Each meal acts as a cue for peripheral clocks in the liver, pancreas, adipose tissue, and gut. These tissues do not simply respond to nutrients; they anticipate them. Regular feeding times reinforce circadian alignment.

When meals occur consistently during daylight hours, metabolic systems synchronize. Insulin sensitivity aligns with glucose arrival. Digestive enzymes activate predictably. Energy flow stabilizes.

When meal timing becomes irregular — skipping breakfast, late dinners, eating under artificial light at night — **phase** begins to drift. Peripheral clocks desynchronize from the central circadian clock. The body receives conflicting signals.

Light may say "day."
Food at midnight says "day."
Melatonin says "night."

This internal disagreement weakens metabolic coherence.

Meal timing is not about rigid schedules. It is about preserving alignment between internal clocks and environmental cues.

Consistent timing strengthens **synchronization**.
Irregular timing fragments it.

Metabolism thrives on predictability.

8.3 Glucose Variability

Blood sugar is often measured as a single number — fasting glucose, A1C, post-meal reading.

But glucose, like all biological systems, moves in waves.

After eating, glucose rises. Insulin responds. Glucose falls. Between meals, levels stabilize. This oscillation is normal and healthy.

The question is not whether glucose rises.

The question is how it rises — and how it falls.

Healthy glucose variability reflects appropriate **amplitude** and efficient **recovery**. The rise is controlled. The fall is complete. Baseline remains stable.

Distorted variability appears in two forms:

- Exaggerated spikes followed by sharp crashes.
- Persistently elevated baseline without clear descent.

Both patterns indicate misalignment between insulin pulses, meal timing, and circadian phase.

Excessive variability can produce fatigue, irritability, and cravings. Reduced variability — a flattened pattern from persistently elevated glucose — signals impaired recovery and rising baseline.

Glucose variability is a window into oscillatory integrity.

When rhythms align, glucose curves remain smooth and predictable. When synchronization falters, the curve becomes chaotic or plateaued.

Metabolic health is not static glucose control.

It is rhythmic stability.

Energy is healthiest when it rises and falls in coherent patterns — not in constant elevation, not in unpredictable spikes, but in coordinated oscillation.

8.4 Circadian Feeding Windows

The body does not process nutrients equally at all hours.

There is a window each day when metabolism is primed — when **insulin sensitivity** is higher, digestive enzymes are active, and peripheral clocks in the liver and pancreas are synchronized with the central circadian clock.

This window is shaped by light.

During daylight hours, especially in the morning and early afternoon, metabolic tissues anticipate feeding. Glucose handling is more efficient. Insulin pulses are clearer. Recovery after meals is faster.

At night, physiology shifts.

Melatonin rises. Body temperature falls. Insulin sensitivity declines. The body prepares for repair, not intake.

When food enters during biological night, the signal conflicts with internal timing. Peripheral clocks attempt to reset. Synchronization weakens. Glucose may remain elevated longer. Insulin recovery slows.

A **circadian feeding window** — a consistent daily interval aligned with daylight — supports phase coherence. It does not require perfection. It requires alignment.

When feeding is consolidated into predictable hours and avoided during late biological night, metabolic oscillators strengthen their coordination.

This is not a diet.

It is timing hygiene.

Circadian feeding preserves:

- Proper **phase alignment**
- Efficient **insulin recovery**
- Stable **baseline glucose**
- Cross-system **synchronization**

Energy stabilizes not only because of what is eaten, but because the body knows when to expect it.

8.5 The Cost of Desynchronization

Desynchronization rarely feels dramatic at first.

It feels like subtle instability.

An afternoon crash.
Late-night hunger.
Morning fog.
Inconsistent energy despite adequate calories.

When **phase** drifts and **synchronization** weakens, metabolic oscillators lose coherence. Insulin pulses misalign with glucose arrival. Peripheral clocks fall out of step with the central clock. Recovery shortens. Baseline gradually shifts.

The cost accumulates quietly.

Elevated fasting glucose.
Reduced insulin sensitivity.
Increased inflammatory tone.
Altered lipid metabolism.

Over months and years, desynchronization contributes to metabolic syndrome, weight gain, and systemic inflammation — not because of one dramatic failure, but because oscillatory integrity has eroded.

The body was designed to anticipate.

When timing becomes unpredictable, anticipation becomes confusion.

Energy is still consumed.
Hormones are still released.
But the waves no longer reinforce one another.

Instead of coordinated rise and fall, there is friction.

The solution is not constant restriction or biochemical force.

It is restored alignment.

Metabolism stabilizes when:

- Light exposure anchors **phase**.
- Feeding aligns with circadian timing.
- Sleep protects **recovery**.
- Oscillators synchronize across systems.

Desynchronization is not merely metabolic imbalance.

It is rhythmic fragmentation.

And as with sleep, emotion, and immunity, restoration does not require eliminating fluctuation.

It requires restoring coherence.

Metabolic health is not about suppressing waves.

It is about letting them move in time.

Chapter 9 — Social Synchrony and Nervous System Stability

The Rhythm Between Us

She could calm herself when alone.

Breathing exercises helped. Meditation helped. Sleep helped — sometimes.

But the deepest calm came unexpectedly.

Sitting across from a friend.
Hearing a steady voice.
Feeling understood.

Her shoulders softened without effort.
Her breathing slowed naturally.
The tension she had been carrying dissolved — not because she forced it to, but because something aligned.

The body had synchronized.

We often think of regulation as an internal skill — something we do within ourselves. But human physiology evolved in connection. The nervous system was not designed to oscillate alone.

The primary variable in social physiology is **synchronization**.

When two people interact safely, their autonomic rhythms begin to align. Breathing patterns subtly entrain. Heart rates shift in coordinated variability. Neural activity mirrors across brains. Emotional tone stabilizes.

This is not metaphor.

It is biology.

The **vagus nerve** links facial expression, vocal tone, and heart rhythm. The cadence of speech influences autonomic state. Eye contact modifies neural oscillations. Touch alters hormonal pulses.

We regulate each other.

When synchronization is intact, stress oscillations shorten. Recovery improves. Immune tone stabilizes. Emotional amplitude remains flexible. Baseline tension lowers.

When synchronization fractures — through isolation, conflict, chronic loneliness — oscillators drift apart. Stress frequency increases. Recovery weakens. Baseline vigilance rises.

Loneliness is not merely psychological pain.

It is rhythmic disruption.

In this chapter, we will explore how **co-regulation** shapes nervous system stability, how **heart rate entrainment** reflects social alignment, how **oxytocin pulses** support connection, how loneliness distorts synchronization, and how belonging stabilizes baseline physiology.

Because the human organism does not only oscillate internally.

It oscillates relationally.

And often, the most powerful medicine is not biochemical suppression or behavioral control.

It is restored synchronization.

By the end of this part, one truth should be unmistakable:

Sleep restores rhythm.
Emotion preserves amplitude.
Immunity depends on pulsatility.

Metabolism follows time.
Connection stabilizes synchronization.

Medicine is not always about eliminating waves.

It is often about restoring them.

Medicine is often rhythm restoration.

9.1 Co-Regulation

From infancy onward, the human nervous system learns rhythm through relationship.

A newborn cannot regulate its own breathing, heart rate variability, or stress cycles efficiently without external input. A caregiver's voice, touch, and facial expression provide timing cues. The infant's nervous system entrains to the adult's steadier rhythm. Over time, internal regulation develops from this external scaffolding.

This process never fully disappears.

As adults, we still experience **co-regulation** — the phenomenon in which one nervous system stabilizes another. A calm presence can reduce **stress amplitude**. A steady voice can slow **respiratory frequency**. A reassuring touch can accelerate **recovery** after activation.

Co-regulation works because oscillators influence one another.

When one system maintains stable variability, nearby systems tend to synchronize. In safe social contact, the autonomic nervous system shifts toward parasympathetic balance. Baseline tension lowers. Emotional oscillations widen without becoming chaotic.

Co-regulation is not weakness.

It is biological design.

Isolation removes this buffering mechanism. Without external rhythmic support, the nervous system must self-stabilize under all conditions. Stress cycles may accelerate. Recovery may shorten. Baseline vigilance may drift upward.

We are not built to regulate alone.

We are built to synchronize.

9.2 Heart Rate Entrainment

The heart is not isolated from social experience.

During meaningful interaction, subtle alignment can occur between individuals' heart rhythms. This phenomenon — **heart rate entrainment** — reflects synchronization of autonomic oscillations.

When two people converse calmly, breathe at similar tempos, or share attention, their **heart rate variability** patterns may begin to align. This does not mean their hearts beat identically. Rather, their oscillatory patterns show coherence.

Breathing influences heart rate through the vagus nerve. When breathing patterns synchronize, cardiac rhythms often follow. Emotional tone, posture, and vocal cadence further shape this alignment.

Entrainment enhances stability.

When heart rhythms remain flexible and coordinated with breathing, **recovery** improves. Stress amplitude decreases more quickly. Baseline autonomic tone stabilizes.

Disrupted social interaction — hostility, unpredictability, chronic conflict — can produce the opposite effect. Oscillations fragment. Variability narrows. Frequency increases.

Heart rate entrainment demonstrates a powerful principle:

Synchronization is measurable.

And it is protective.

The body does not regulate in isolation; it calibrates through shared rhythm.

9.3 Oxytocin Pulses

Oxytocin is often called the "bonding hormone," but like other signaling molecules, it functions in pulses.

Oxytocin release increases during social touch, eye contact, trust-building interactions, and moments of emotional safety. These pulses influence autonomic balance, reduce stress reactivity, and support social memory formation.

Importantly, oxytocin does not remain elevated continuously. It rises in response to connection and falls afterward. Its oscillatory nature allows responsiveness without overstimulation.

When oxytocin pulses occur in healthy patterns, they contribute to:

- Lowered **stress frequency**
- Improved **recovery** after activation
- Stabilized emotional **amplitude**
- Reduced inflammatory signaling

In contrast, chronic isolation may reduce oxytocin pulsatility. Stress systems may become less buffered. Baseline tension may remain elevated.

Oxytocin illustrates that connection is not abstract.

It is rhythmic chemistry.

Social interaction shapes physiological oscillation through timed pulses — not constant states.

Connection is not merely psychological nourishment.

It is biological synchronization.

And when synchronization is preserved, nervous system stability follows.

9.4 Loneliness as Rhythm Disruption

Loneliness is often described as an emotion.

But physiologically, it behaves like a state of misalignment.

When social synchronization diminishes, the nervous system loses an external stabilizer. The buffering effects of **co-regulation** weaken. Stress cycles lengthen. **Recovery** slows. **Baseline vigilance** rises.

Research shows that chronic loneliness is associated with elevated inflammatory markers, altered cortisol rhythms, and reduced heart rate variability. These are not random correlations. They reflect distortion of key oscillatory variables:

- **Frequency** of stress activation increases.
- **Recovery** becomes incomplete.
- **Synchronization** between systems weakens.
- **Baseline** inflammatory tone drifts upward.

Loneliness is not merely the absence of company. It is the absence of rhythmic alignment.

Without social entrainment, internal oscillators must self-regulate under all conditions. The body remains slightly guarded. Hypervigilance becomes subtle but persistent. Sleep may fragment. Immune pulsatility may narrow.

The wave still exists.

But it loses coherence.

Over time, this desynchronization can influence mood, energy regulation, immune tone, and even metabolic stability. The organism begins to operate in partial isolation — a state evolution never designed us to endure for long.

Loneliness is rhythm without resonance.

9.5 Belonging as Baseline Stabilizer

Belonging does more than provide comfort.

It stabilizes **baseline**.

When an individual feels securely connected, the nervous system recalibrates. Cortisol rhythms normalize. Inflammatory markers decline. Heart rate variability improves. Emotional amplitude becomes more flexible without tipping into volatility.

Belonging lowers the resting line.

It does not eliminate stress. It improves **recovery**.
It does not suppress emotion. It preserves **range**.
It does not remove challenge. It strengthens **synchronization**.

From an oscillatory perspective, belonging acts as a stabilizing field. It reduces chronic elevation of stress signals and supports efficient descent after activation.

This is why social rhythm therapies — consistent routines, shared activities, predictable interaction — often improve mood and energy regulation. They restore timing and reinforce synchronization.

Belonging is not an accessory to health.

It is part of its architecture.

The body stabilizes most effectively when it does not oscillate alone.

End of Part II. Restoration Comes Through Realignment, Not Suppression

Across sleep, emotion, immunity, metabolism, and connection, one principle has repeated itself:

Health is not the suppression of fluctuation.

It is the preservation of adaptive oscillation.

Sleep restores **phase** and **recovery**.
Emotion preserves **amplitude**.
Immunity depends on controlled **frequency** and adaptive **baseline**.
Metabolism requires **synchronization** with time.
Connection stabilizes physiological rhythm.

Distortion in these variables produces symptoms.
Restoration of rhythm restores coherence.

Medicine is not always about silencing waves.

It is often about realigning them.

Medicine is often rhythm restoration.

When Flexibility Narrows

Before we move from restoration to illness, one clarification matters.

Illness is not the opposite of rhythm.
It is rhythm that has lost flexibility long enough to become structure.

A healthy system rises and falls within adaptive bounds.
It activates and resolves.
It expands and returns.

Distortion rarely begins with collapse. It begins with narrowing.

Recovery shortens. Baselines drift. Phase misaligns.
Synchronization weakens.

At first, the wave still moves.
But its range decreases.
Its descent becomes incomplete.

Over time, what was once oscillation stabilizes into rigidity.

And rigidity, when sustained, reshapes physiology.

Illness does not emerge because rhythm disappears.
It emerges because rhythm can no longer modulate freely.

The difference between medicine and illness is not the presence of fluctuation.

It is whether flexibility remains intact.

When flexibility remains, restoration is possible.
When flexibility narrows, distortion accumulates.
When flexibility collapses, pattern consolidates into pathology.

Health depends on movement within bounds.
Illness reflects the loss of that movement.

Glossary — Part II: Rhythm as Medicine

Adaptive Amplitude
The healthy expansion and contraction of emotional, physiological, or metabolic states.
Adaptive amplitude allows strong activation followed by complete recovery without volatility.

Belonging
A state of secure social connection that stabilizes physiological baseline, improves stress recovery, and supports autonomic synchronization.

Circadian Feeding Window
A consistent daily eating interval aligned with daylight hours that strengthens metabolic phase alignment and insulin pulsatility.

Co-Regulation
The process by which one nervous system stabilizes another through social interaction, voice tone, facial expression, touch, and shared rhythm.

Controlled Oscillation
A rise-and-fall pattern that is strong but contained, allowing activation without chronic elevation.

Deep (Slow-Wave) Sleep
A stage of sleep characterized by high-amplitude delta brain waves that support tissue repair, immune recalibration, hormonal release, and glymphatic clearance.

Desynchronization
Loss of coordination between biological rhythms (e.g., sleep, metabolism, immune function), leading to fragmented oscillatory integrity.

Emotional Rebound
The natural descent following emotional or dopamine elevation, necessary to preserve sensitivity and future amplitude.

Entrainment
The alignment of internal biological rhythms with external cues such as light, social timing, or environmental structure.

Fever (Controlled Immune Oscillation)

A temporary upward shift in thermal baseline that enhances immune defense and resolves after threat reduction.

Glymphatic Clearance
The sleep-dependent movement of cerebrospinal fluid that removes metabolic waste from the brain, strongest during deep oscillatory sleep.

Immune Pulsatility
The rhythmic rise and fall of immune activation, allowing effective defense followed by resolution and return to baseline.

Inflammatory Baseline
The resting level of immune activation.
A low adaptive baseline allows strong acute response; a shifted baseline reflects chronic activation.

Insulin Pulsatility
The rhythmic release of insulin in discrete bursts rather than constant elevation, preserving cellular sensitivity and metabolic health.

Metabolic Synchronization
Alignment of feeding behavior, insulin release, circadian timing, and cellular metabolic clocks.

Oscillatory Restoration
Therapeutic recalibration achieved not by eliminating fluctuation, but by restoring proper rise, fall, alignment, and recovery.

Phase Alignment
The synchronization of internal rhythms with environmental time cues, especially light and feeding cycles.

Recovery Integrity
The completeness and efficiency of return to baseline after activation, especially during sleep.

Social Synchronization
The physiological alignment of autonomic rhythms between individuals during safe interaction.

Stress Rebound
The recovery phase following activation, essential for resilience and prevention of baseline elevation.

Synchronization
The coordinated timing of multiple oscillators within the body, allowing systems to rise and fall together.

Timing Hygiene
Daily practices that preserve rhythmic integrity (consistent sleep, aligned light exposure, structured feeding).

Wave Restoration
The principle that many forms of healing occur through re-establishing rhythmic rise and fall rather than suppressing biological activity.

PART III — Rhythm as Illness

When the Pattern Distorts

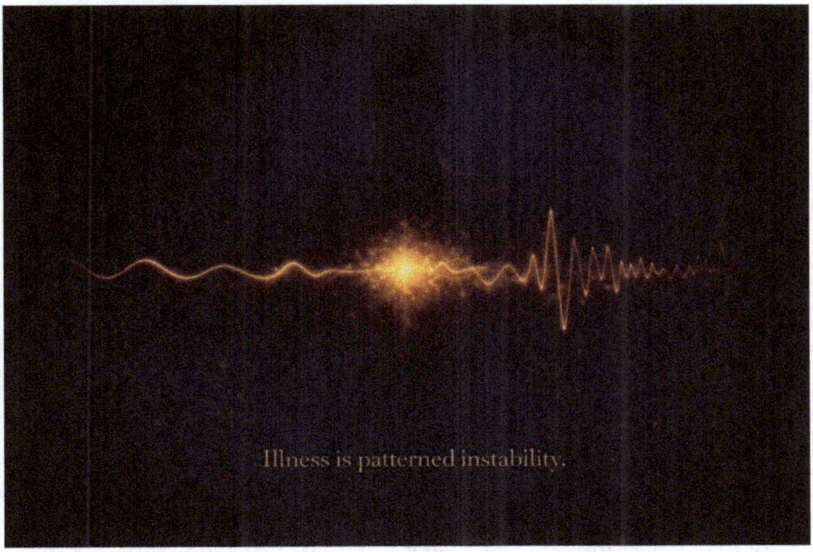

By now, the language of oscillation is familiar.

Health moves through **amplitude**, **frequency**, **phase**, **synchronization**, **recovery**, and **baseline**. When these variables remain flexible and coordinated, vitality follows.

Illness often does not begin with sudden collapse.

It begins with gradual distortion.

Amplitude narrows — or overshoots.
Frequency accelerates — or compresses.
Phase drifts out of alignment.
Recovery shortens.
Baseline shifts upward — or downward.
Synchronization weakens across systems.

The wave does not disappear. It continues to move — but with reduced flexibility.

At first, the change may be subtle.

A little less contrast in mood.
A little more background tension.
Sleep slightly mistimed.
Stress cycles slightly faster.
Inflammation resolving more slowly.

Nothing catastrophic.

But the system no longer regulates with ease.

This section does not claim that all disease reduces to rhythm. Biology is shaped by genetics, trauma, environment, structural pathology, and countless interacting variables.

What this framework offers is a lens.

A way to recognize that many disorders reflect rigid or maladaptive alterations in oscillatory variables.

Depression may involve flattened **amplitude**.
Bipolar states may reflect excessive **amplitude**.
Anxiety may reflect accelerated **frequency**.
Insomnia may involve **phase** delay.
Burnout may reflect impaired **recovery** and weakened **synchronization**.
Chronic inflammation may signal an elevated **baseline**.

When we identify which variable is under strain, intervention becomes clearer.

Because disorder is often not moral failure. It is pattern distortion.

And when the pattern is understood, flexibility can begin to return.

Figure 3.0 — Mapping Disorders to Oscillatory Distortion. *Many common patterns of illness can be understood as distortions in one or more oscillatory variables. The table below summarizes primary and secondary variables often involved.*

Pattern Distortion Mapping

Condition / Pattern	Primary Variable Distortion	Secondary Variables Often Involved
Depression	↓ Amplitude	↓ Recovery, ↑ Baseline
Anhedonia	↓ Amplitude	↓ Frequency
Chronic Fatigue	↓ Amplitude	↓ Recovery, ↑ Baseline
Bipolar Spectrum	↑ Amplitude	↓ Recovery, Phase Instability
Emotional Volatility	↑ Amplitude	↑ Recovery, ↑ Baseline
Anxiety Disorders	↑ Frequency	↑ Amplitude
Panic Disorder	↑ Frequency (Acute Spikes)	↑ Recovery, Phase Instability
Insomnia	Phase Delay / Phase Instability	↓ Recovery
Shift Work Disorder	Phase Misalignment	Desynchronization
Burnout	↓ Recovery	↑ Baseline, Desynchronization
Chronic Inflammation	↑ Baseline	↓ Recovery
Metabolic Syndrome	Desynchronization	↑ Baseline
Hypervigilance	↑ Frequency	↑ Baseline
Emotional Numbing	↓ Amplitude	↑ Baseline Rigidity

Figure 3.0 Pattern-level mapping of common conditions to primary oscillatory distortions. *This framework does not reduce illness to a single variable but highlights how persistent dysregulation in amplitude, frequency, phase, synchronization, recovery, or baseline can contribute to recognizable patterns across mood, sleep, metabolism, stress, and immune function.*

The chapters that follow examine each distortion category in depth. This mapping is not diagnostic. It is structural — a way to identify which variable may be under strain so that restoration can become targeted rather than generalized.

Chapter 10 — Flattened Amplitude

When the Wave Loses Height

He did not feel devastated.
He did not feel panicked.
He did not even feel particularly sad.

He felt... muted.

Music sounded thinner.
Food tasted ordinary.
Success registered intellectually but not emotionally.
Laughter required effort.

Nothing catastrophic had happened. Nothing visibly broken.

But the highs were no longer high.
And the lows were no longer deep.

The emotional landscape had flattened.

What has narrowed here is **amplitude** — the height of emotional rise and fall.

When amplitude compresses, contrast fades. Peaks soften. Motivation dulls. The system dampens variation, often as a form of protection — but at the cost of vitality.

This is not the same as sadness.

Sadness is a descent within a healthy oscillation. It carries depth, meaning, and the capacity to return.

Flattened amplitude feels different.

It feels like reduced signal.

Emotion becomes shallow. Motivation weakens. Energy dulls. The nervous system continues to move — but with diminished range.

In many cases, depression reflects this narrowing. Dopamine pulses lose intensity. Reward circuitry responds less robustly. Engagement no longer generates strong internal rise.

The same pattern appears in chronic fatigue. Energy does not surge even after rest. Effort feels heavier. Activation never quite reaches lift.

Flattened amplitude is not laziness.
It is not ingratitude.
It is not moral weakness.

It is a change in the height of the wave.

In this chapter, we examine how depression, anhedonia, chronic fatigue, and emotional narrowing can be understood — in part — as distortions of **amplitude**.

Not to reduce their complexity.

But to clarify their structure.

Because when amplitude narrows, life does not disappear.

It quiets.

And recognizing the quieting is the first step toward restoring range.

10.1 Depression and Blunted Reward

Depression is often described as sadness.

But many people with depression will tell you something different.

"I don't feel sad," they say.
"I feel empty."
"I feel nothing."

This distinction matters.

Sadness is an emotional descent within a healthy oscillation. It carries tone, texture, and narrative. It moves in response to loss and, over time, returns.

Depression, in many cases, reflects something else: a reduction in **amplitude**.

The nervous system still cycles. Circadian rhythms continue. Stress hormones rise and fall. But the emotional peaks flatten. The rise of motivation weakens. Reward signals diminish.

One of the core neurobiological features often associated with depression is altered **dopamine signaling** in reward pathways. When anticipated rewards fail to generate strong dopamine pulses, the experience of engagement dulls. Activities that once felt meaningful now register faintly.

This is sometimes called **blunted reward responsiveness**.

In oscillatory language, the wave still exists — but it has lost height.

It is essential to state clearly that depression is not reducible to a single variable. Major depressive disorder involves genetic vulnerability, structural and functional alterations in fronto-limbic circuitry, inflammatory and endocrine influences, developmental history, and trauma exposure. In some individuals, amplitude narrowing may be primary. In others, it may be secondary to immune activation, thyroid dysfunction, circadian disruption, or neurobiological injury. The oscillatory lens does not replace these models. It offers a structural pattern within which they may operate.

The brain may adapt this way under chronic stress, inflammation, or prolonged adversity. Dampening amplitude can be protective in extreme environments. High peaks are metabolically expensive. Emotional intensity requires energy.

But when amplitude remains chronically narrowed, vitality suffers.

The individual may still function. They may meet obligations, complete tasks, maintain routines. Yet the inner contrast — the dynamic rise that makes effort feel worthwhile — weakens.

Understanding depression through the lens of amplitude does not deny its complexity. It does not replace genetics, trauma, or social context.

It highlights a pattern:

When emotional amplitude narrows and recovery falters, reward feels distant.

The problem is not that the person cannot try harder.

It is that the wave has lost height.

10.2 Anhedonia

Anhedonia is often defined simply as the inability to feel pleasure.

But that definition misses something essential.

Anhedonia is not only about pleasure.

It is about anticipation.

In healthy oscillation, dopamine pulses rise before reward — during expectation. The promise of engagement fuels motivation. The brain predicts satisfaction and prepares the body for action.

Anhedonia disrupts this predictive rise.

The individual may still understand that an activity is enjoyable in theory. They may remember that they once loved music, conversation, or movement. But the anticipatory spark does not ignite. The upward surge fails to form.

In oscillatory terms, the amplitude of the reward wave is reduced before it even peaks.

Without anticipatory rise, behavior loses momentum. Effort feels heavier because the nervous system no longer signals upcoming contrast.

Importantly, anhedonia does not always mean complete absence of feeling. It often manifests as muted pleasure — diminished rather than eliminated.

This narrowing can extend beyond joy.

Curiosity dulls.
Interest softens.
Emotional highs compress toward neutrality.

Over time, when amplitude remains flattened, baseline mood may drift downward. The nervous system becomes accustomed to operating within a narrower range.

Anhedonia illustrates a critical principle of the Oscillatory Health Model:

Health requires contrast.

Without sufficient amplitude, life becomes grayscale.

But recognizing this pattern reframes the experience.

It is not that pleasure has vanished.

It is that the wave has narrowed.

And if amplitude can narrow, it can also expand.

The task is not to force joy.

It is to restore oscillation.

10.3 Chronic Fatigue

Chronic fatigue is often mistaken for simple tiredness.

But tiredness resolves with rest.
Chronic fatigue lingers.

In many individuals, sleep duration may be adequate. Laboratory values may appear "normal." Yet energy never fully rises. Mornings feel heavy. Effort feels disproportionately costly. Recovery after exertion takes longer than expected.

From an oscillatory perspective, chronic fatigue frequently reflects reduced **amplitude** in the energy system.

In healthy physiology, energy follows a daily arc. Cortisol rises in the morning. Alertness increases. Metabolic activity mobilizes. Throughout the day, activity peaks and falls in predictable cycles. After exertion, recovery restores baseline.

When amplitude narrows, that arc flattens.

Morning activation may be blunted.
Afternoon energy may not rebound.
Exercise may fail to generate invigorating contrast.

The rise is weak. The fall is incomplete.

Over time, impaired **recovery** compounds the problem. Each day begins from partial restoration. Baseline energy drifts downward. The system conserves resources by reducing variation.

Importantly, this pattern does not imply laziness or lack of discipline. It often reflects complex interactions between stress physiology, immune signaling, sleep architecture, and metabolic timing.

The organism adapts by dampening amplitude.

But the cost is vitality.

Chronic fatigue illustrates that energy health is not defined by how much one pushes through.

It is defined by the integrity of the rise-and-return cycle.

10.4 Emotional Narrowing

Emotional narrowing is subtler than depression.

The person still feels — but within a reduced range.

Emotional narrowing often presents as reduced affective variability across domains. Individuals may report that joy feels present but attenuated, frustration less reactive yet less energizing, and sadness less intense yet less cathartic. The emotional system remains functional, but operates within a compressed bandwidth.

Life becomes manageable, but less vivid.

Sometimes emotional narrowing develops as a protective adaptation. After trauma or prolonged stress, the nervous system may reduce amplitude to avoid overwhelming peaks. By flattening highs, it also softens lows.

This strategy can preserve function in the short term.

But over time, narrowed amplitude limits connection, creativity, motivation, and meaning.

In oscillatory terms, the wave becomes smaller — not absent, but compressed.

Contrast diminishes.

Emotional narrowing can coexist with productivity. A person may appear stable, composed, even calm. Yet internally, the richness of experience has thinned.

Healthy amplitude allows both expansion and contraction.

Without sufficient range, resilience weakens. The system loses the capacity to mobilize fully when needed and to experience reward deeply.

Emotional health does not require constant intensity.

But it does require range.

Flattened amplitude is not moral failure.
It is adaptive narrowing that has persisted too long.

And as with other oscillatory distortions, recognizing the pattern opens the possibility of restoration.

Because amplitude is not fixed.

It can expand again.

Chapter 11 — Excess Amplitude

When the Wave Rises Too High

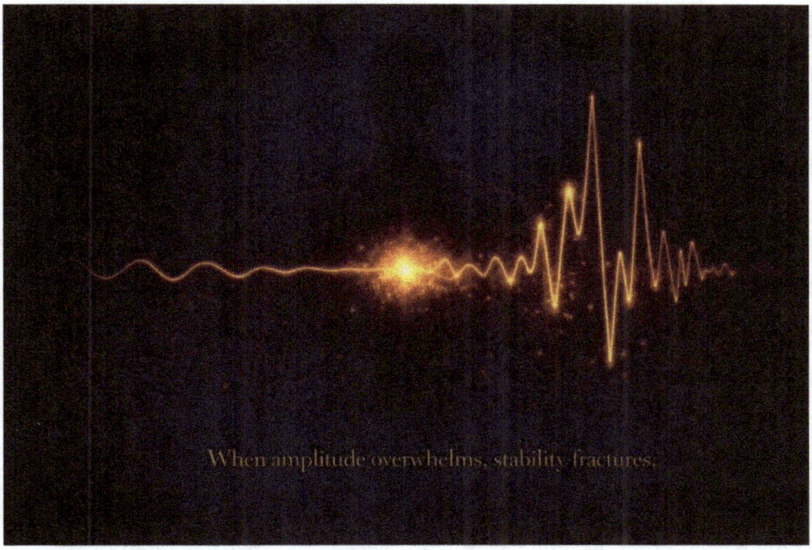

Excess amplitude represents a different category of distortion. Rather than narrowing of contrast, the system generates peaks that exceed regulatory capacity. Emotional, behavioral, and physiological activation intensify beyond what recovery mechanisms can contain.

She did not feel muted.

She felt electrified.

Ideas arrived rapidly. Sleep felt unnecessary. Energy surged through her body like current. Confidence expanded. Projects multiplied. The world seemed brighter, faster, more charged.

And then, just as powerfully, the descent came.

Exhaustion.
Irritability.
Regret.
A crash that felt disproportionate to the ascent.

This was not ordinary emotional fluctuation.

It was amplified oscillation.

In the language of the Oscillatory Health Model, this is distortion of **amplitude** in the opposite direction.

Amplitude is necessary for vitality. Without strong rise, motivation dulls. Without emotional depth, meaning thins.

But when amplitude exceeds regulatory capacity, stability fractures.

The nervous system becomes overdriven. Dopamine surges escalate beyond adaptive range. Sleep shortens. Recovery narrows. Baseline may shift upward. The following descent becomes steeper because the preceding peak was too high.

This pattern appears in varying degrees — from everyday emotional volatility to the clinical extremes of the **bipolar spectrum**.

It is essential to be clear:

High energy is not illness.
Excitement is not pathology.
Ambition is not instability.

The distortion emerges when amplitude exceeds control — when peaks compromise recovery and troughs deepen beyond proportion.

Healthy oscillation allows strong rise with intact return.

Excess amplitude produces rise without adequate regulation.

In this chapter, we will explore how excessive amplitude manifests — from bipolar mood shifts to stress spikes and emotional volatility — and how oscillation becomes destabilizing when modulation fails.

Because amplitude is not the enemy.

Loss of modulation is.

And when the wave rises too high, the system must either regain control — or collapse into compensatory descent.

Excess is not always strength.

Sometimes it is dysregulated rhythm.

Understanding that distinction is the beginning of balance.

11.1 Bipolar Spectrum

Few conditions illustrate distorted **amplitude** as clearly as the **bipolar spectrum**.

At one pole, mood and energy surge beyond typical range. Sleep decreases without perceived fatigue. Ideas accelerate. Confidence expands. Speech quickens. Risk tolerance rises. The world feels charged with possibility.

At the other pole, descent can be equally powerful. Energy collapses. Motivation fades. Sleep may fragment or overshoot. Emotional heaviness deepens.

This is not ordinary emotional fluctuation.

It is amplified oscillation.

From an oscillatory perspective, bipolar patterns often involve:

- Exaggerated **amplitude** of mood and energy
- Impaired **recovery** after peaks
- Possible shifts in **baseline** across episodes
- Disturbances in **phase**, especially sleep timing

During elevated states, dopamine signaling may intensify. Sleep reduction further destabilizes regulatory circuits. Recovery mechanisms weaken as activation persists. The following descent may partly represent the system's attempt to recalibrate after excessive rise.

It is crucial to avoid oversimplification.

Bipolar disorder involves complex genetic, neurochemical, and environmental influences. It cannot be reduced to a single variable.

Yet the lens of amplitude offers clarity.

The problem is not that emotion rises.

It is that the rise exceeds regulatory containment.

Bipolar spectrum disorders are among the most biologically complex mood conditions described in psychiatry. Strong heritable components, ion channel regulation, circadian gene expression, dopaminergic and glutamatergic modulation, and structural network connectivity all contribute to their expression. Excess **amplitude** is a useful descriptive lens — but it does not exhaust causation. The oscillatory model highlights dynamic behavior of the system; it does not substitute for neurobiological reality.

Healthy amplitude allows strong emotion with intact boundaries. In bipolar spectrum conditions, amplitude becomes dysregulated — peaks outrun control, and troughs deepen accordingly.

Understanding this pattern reframes experience.

The individual is not "too much."

The oscillation is exceeding modulation.

Treatment, therefore, often focuses — directly or indirectly — on stabilizing amplitude, protecting sleep phase, and strengthening recovery capacity.

Because the goal is not flattening.

It is controlled oscillation.

11.2 Emotional Volatility

Not all excessive amplitude reaches clinical extremes.

Many people experience **emotional volatility** — rapid shifts in mood intensity that feel disproportionate to events.

Excitement becomes euphoria.
Frustration becomes anger.
Disappointment becomes despair.

The oscillation is fast and high.

In volatility, amplitude may not remain elevated for days or weeks as in bipolar states. Instead, peaks are sharp and short-lived. Frequency may also increase, producing rapid cycling between emotional states.

Several mechanisms can contribute:

- Heightened stress reactivity
- Reduced inhibitory regulation in emotional circuits
- Sleep disruption affecting phase stability
- Chronic inflammation altering mood signaling

The common pattern remains distortion of amplitude.

When emotional peaks rise too high, the nervous system expends energy rapidly. Recovery becomes strained. Baseline may temporarily elevate or drop following intense swings.

Volatility often feels chaotic because the system cannot modulate height effectively.

Yet it is important to distinguish intensity from pathology.

High amplitude alone is not dysfunction.

Intensity becomes destabilizing when:

- **Recovery** fails
- **Synchronization** with sleep and stress rhythms weakens
- Peaks trigger impulsive behavior
- Troughs impair functioning

Emotional volatility represents oscillation without sufficient regulation.

The solution is not emotional suppression.

It is strengthening modulation — protecting sleep, stabilizing stress cycles, preserving recovery.

Because amplitude is meant to give life color.

But without control, the color becomes overwhelming.

Resilience lies not in eliminating intensity.

It lies in ensuring that the wave rises powerfully — and still knows how to return.

11.3 Stress Spikes

Not all excessive amplitude expresses itself through mood.

Sometimes it appears as sudden surges of **stress activation** — sharp physiological spikes that seem disproportionate to the trigger.

A brief email provokes racing heart.
A minor disagreement ignites intense agitation.
A small uncertainty produces a surge of adrenaline.

The stress response is designed for amplitude. It must mobilize energy rapidly in the face of danger. But in healthy oscillation, the spike is followed by efficient **recovery**.

When amplitude becomes exaggerated, stress spikes overshoot.

Heart rate accelerates dramatically.
Cortisol surges strongly.
Muscles tense intensely.
Attention narrows abruptly.

Then, depending on recovery capacity, one of two patterns may follow:

- A clean descent back to baseline.
- Or prolonged activation, where the spike lingers.

Repeated exaggerated spikes can alter **baseline** tone. The system begins to anticipate threat. Frequency of activation increases. Even neutral cues can trigger large amplitude responses.

Stress spikes often reflect heightened sensitivity in autonomic circuits — particularly when sleep phase is unstable, inflammation is elevated, or emotional regulation networks are fatigued.

Again, the issue is not stress itself.

It is disproportion.

A wave that rises too high for the context drains more energy than necessary. The body compensates through fatigue, irritability, or emotional collapse.

Healthy amplitude mobilizes.
Excess amplitude overwhelms.

11.4 Oscillation Without Control

When amplitude becomes excessive across systems — mood, stress, energy, sleep — oscillation begins to lose modulation.

The wave still rises and falls.

But the governing structure weakens.

Peaks override judgment.
Descents impair function.
Sleep shortens during highs and fragments during lows.
Recovery fails to stabilize the cycle.

This is **oscillation without control**.

Control does not mean suppression. It means containment — the ability to generate strong amplitude within safe bounds.

Without containment, oscillation becomes destabilizing rather than adaptive.

Several factors can erode control:

- Chronic sleep disruption affecting **phase**
- Repeated stress impairing **recovery**
- Inflammatory shifts altering **baseline**
- Genetic vulnerabilities affecting regulatory circuits

As control weakens, amplitude may become increasingly self-amplifying. High states shorten sleep, which further destabilizes regulation. Low states impair energy, which weakens resilience.

The system oscillates — but without sufficient boundaries.

The lesson is subtle but essential:

Amplitude is not the enemy.

Amplitude gives vitality, creativity, and motivation.

But amplitude must remain modulated by intact **phase**, strong **recovery**, and stable **baseline**.

When modulation fails, oscillation destabilizes.

The goal is not to flatten the wave.

It is to restore control over its height.

Because health requires not small waves —

But waves that rise boldly and return safely.

Chapter 12 — Accelerated Frequency

When the Wave Moves Too Fast

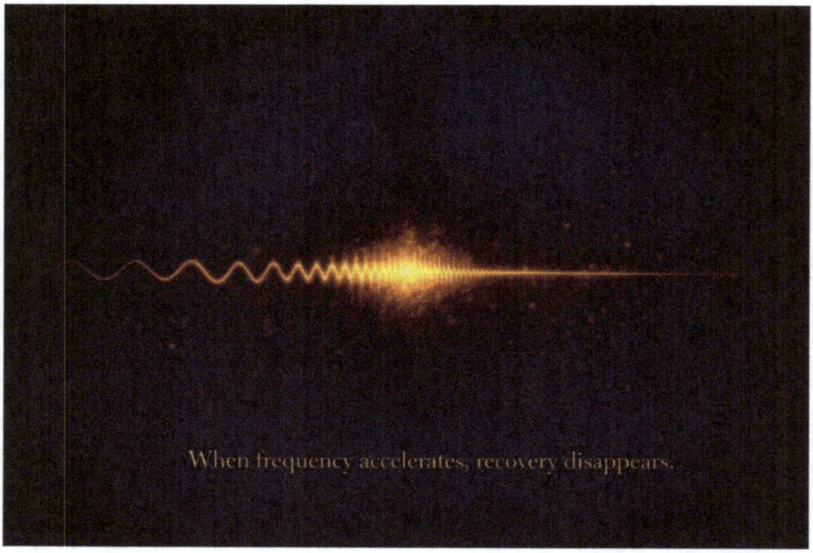

Not all suffering comes from how high the wave rises.

Sometimes the problem is how fast it moves.

Many individuals experiencing chronic anxiety do not report extreme mood swings or overt crisis. Instead, they describe persistent internal acceleration — thoughts cycling rapidly, physiological tension lingering between events, and difficulty achieving full descent.

Thoughts cycled rapidly.
Heart rate lifted easily.
Small uncertainties triggered immediate internal rehearsal.
Even during rest, something hummed beneath the surface.

Nothing catastrophic. Just speed.

In oscillatory terms, this is distortion of **frequency**.

Frequency describes how often a cycle repeats. In healthy systems, activation occurs when needed, then quiets. Stress rises in response to challenge and resolves. Attention narrows briefly and widens again.

When frequency accelerates, cycles repeat before recovery completes.

Activation overlaps with activation.
Alertness persists.
The system barely descends before rising again.

Anxiety, panic, hypervigilance, and chronic stress states often share this pattern: increased cycling of arousal.

The wave is not necessarily higher than normal.
It is simply arriving too often.

This distinction matters.

In accelerated frequency states, the nervous system remains in repeated micro-activation. The body anticipates threat rapidly. Thoughts loop quickly. Cortisol pulses shorten. Recovery compresses.

Over time, persistent acceleration can narrow amplitude, elevate baseline, and fragment sleep phase. The oscillators lose spacing.

In this chapter, we will explore how **anxiety cycles**, **panic loops**, **hypervigilance**, and **chronic stress oscillation** reflect increased frequency — not moral weakness, not lack of resilience, but altered temporal pacing.

Because sometimes the problem is not that the wave is too large.

It is that it never slows down.

And health, as we have seen, depends not only on height —

But on timing.

When oscillation accelerates beyond sustainable pace, the system exhausts itself.

Restoration begins not by eliminating the wave —

But by slowing its tempo.

12.1 Anxiety Cycles

Anxiety is often described as fear without clear danger.

But physiologically, anxiety frequently reflects accelerated **frequency** of arousal cycles.

In a healthy stress response, activation occurs in proportion to threat. Heart rate rises. Attention sharpens. Cortisol pulses. Then, once the situation resolves, the system descends.

In anxiety states, activation cycles repeat more rapidly.

A thought triggers a stress response.
Before recovery completes, another thought triggers another.
Physiological arousal overlaps itself.

The amplitude may not be extreme. The person may still function, speak, work, and sleep — though lightly. But internally, the oscillations are tightly spaced.

Short rise.

Short fall.

Rise again.

This repeated cycling reduces recovery time. Baseline tension gradually elevates. The nervous system becomes more sensitive to small cues. Even ambiguous signals can trigger activation.

Anxiety cycles often involve:

- Increased **frequency** of stress activation
- Shortened **recovery**
- Mild elevation of **baseline vigilance**
- Sleep phase vulnerability

Importantly, anxiety is not necessarily a failure of logic.

It is often a temporal distortion.

Anxiety disorders encompass trauma-related syndromes, genetic predispositions in threat circuitry, amygdala hypersensitivity, HPA-axis dysregulation, autonomic conditioning, and endocrine influences. In some individuals, accelerated **frequency** reflects learned hypervigilance after trauma. In others, it reflects inherited sensitivity or inflammatory priming. The oscillatory framing emphasizes pacing of activation, not a dismissal of underlying neurobiology.

The mind may generate plausible concerns. The body responds as if each requires immediate action. The pacing is the problem.

When frequency slows — through breath pacing, improved sleep phase, social synchronization, or reduced inflammatory tone — anxiety intensity often decreases even if external stressors remain.

The wave does not need to disappear.

It needs space.

12.2 Panic Loops

If anxiety cycles represent rapid repetition, panic represents rapid escalation.

In a panic episode, the stress response accelerates abruptly. Heart rate surges. Breathing quickens. Dizziness, chest tightness, and derealization may appear. The body interprets internal signals as threat, amplifying activation further.

This creates a **loop**.

Physical sensation → threat interpretation → increased activation → intensified sensation → further threat interpretation.

The oscillation does not merely repeat faster; it self-amplifies.

Frequency increases dramatically. Recovery is nearly absent until the episode peaks and exhaustion forces descent.

Panic loops illustrate how frequency and amplitude can interact. The rapid cycling of arousal increases amplitude temporarily, which further accelerates cycling.

Yet even here, the wave eventually falls.

The episode resolves. Heart rate slows. Breathing normalizes. Baseline returns — often with fatigue.

Panic does not reflect broken physiology.

It reflects runaway oscillation.

The system misreads internal cues and accelerates beyond regulatory control.

Understanding panic as accelerated frequency changes the narrative.

The person is not fragile.

The system is mis-timed.

Interventions that slow breathing, anchor attention, and restore recovery spacing directly target frequency. They create temporal gaps that allow the oscillation to descend.

Panic is not infinite.

It is fast.

And when the tempo slows, stability begins to return.

12.3 Hypervigilance

Hypervigilance is not always dramatic.

It is not necessarily panic or visible anxiety.

It is the feeling of never fully powering down.

The mind scans.
The ears stay tuned.
The body remains slightly braced.

Even in quiet environments, something inside remains alert.

In oscillatory terms, hypervigilance reflects persistently accelerated **frequency** of low-level stress activation.

The wave may not spike high.
But it rises often.

Small sounds trigger micro-surges of arousal. Subtle changes in tone provoke internal recalculation. The nervous system cycles through brief activation bursts repeatedly across the day.

Because recovery windows are short, the system never fully descends.

Over time:

- **Baseline vigilance** drifts upward.
- **Recovery efficiency** decreases.
- Sleep phase becomes vulnerable.
- Emotional amplitude narrows from fatigue.

Hypervigilance is common after trauma, chronic stress exposure, or prolonged uncertainty. The nervous system learns that threat may appear at any time, so it increases scanning frequency.

The body adapts by accelerating internal cycles.

The cost is exhaustion.

Hypervigilance is not weakness.

It is an over-trained alarm system.

And the distortion lies not in the existence of vigilance — but in its pacing.

12.4 Chronic Stress Oscillation

When stress cycles accelerate for weeks, months, or years, oscillation changes character.

Instead of distinct activation followed by clear descent, the system enters a pattern of near-continuous micro-activation.

Cortisol pulses become irregular.
Heart rate variability narrows.
Sleep deep cycles shorten.
Inflammatory tone rises.

The wave still exists — but its rhythm becomes compressed.

In healthy stress oscillation:

Activation → Peak → Recovery → Baseline.

In chronic stress oscillation:

Activation → Partial descent → Activation → Partial descent → Activation.

The system rarely touches true baseline.

Frequency remains high. Recovery remains incomplete.

Eventually, amplitude may also narrow. The nervous system dampens emotional peaks to conserve energy. Motivation dulls. Fatigue accumulates.

Chronic stress oscillation demonstrates how variables interact:

Accelerated **frequency** erodes **recovery**.
Impaired recovery elevates **baseline**.
Elevated baseline reduces adaptive **amplitude**.

The system becomes rigid through repetition.

Yet even here, the framework offers hope.

If distortion lies in pacing, intervention lies in restoring temporal spacing:

- Slowing breath to reduce frequency.
- Protecting sleep to restore recovery.
- Reinforcing social synchronization.
- Reducing inflammatory load.

Chronic stress is not proof of fragility.

It is the outcome of sustained acceleration without sufficient descent.

The solution is not elimination of stress.

It is restoration of rhythm.

Because when oscillation regains space — when activation and recovery are allowed to alternate fully — stability begins to return.

And with it, resilience.

Chapter 13 — Phase Delay

When the Wave Falls Out of Time

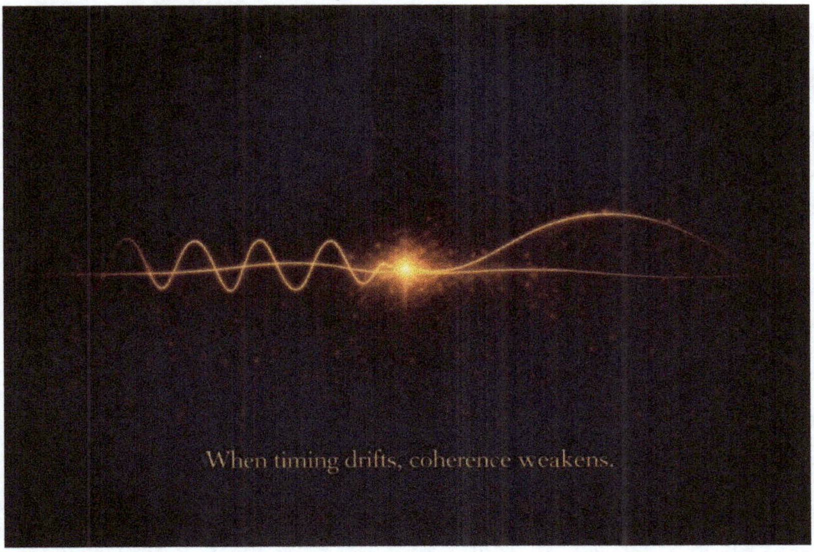

He was exhausted at midnight — but wired.

Sleep felt close, yet unreachable. His mind slowed, but his body resisted descent. When he finally drifted off, morning came too quickly. The alarm felt violent. Cortisol had not yet risen. Light felt intrusive.

At night, he was alert.
In the morning, he was fogged.

Nothing was broken.

But everything was mistimed.

This is distortion of **phase**.

Phase does not describe how high a wave rises or how often it cycles.

It describes *when* it occurs.

In healthy oscillation, internal rhythms align with environmental cues. Light anchors wakefulness. Darkness supports sleep. Cortisol rises before waking. Melatonin peaks at night. Insulin sensitivity aligns with feeding. Immune recalibration occurs during deep sleep.

When phase shifts — even subtly — systems fall out of alignment.

Sleep may begin later than intended.
Hormones may peak at the wrong time.
Energy may surge at night and fade during the day.

The oscillation still exists.

But it is misaligned with the world.

Modern life exerts powerful pressure on phase: artificial light at night, inconsistent schedules, social obligations, shift work, travel across time zones. The internal clock, designed for solar rhythm, must negotiate a 24-hour artificial environment.

The consequences accumulate.

Sleep fragments.
Cortisol timing shifts.
Inflammatory tone alters.
Metabolic regulation weakens.

Phase delay does not require dramatic insomnia to cause harm.

Even modest misalignment can distort recovery and synchronization across systems.

In this chapter, we will explore how **insomnia**, **shift work disorder**, **hormonal misalignment**, and **metabolic drift** reflect phase distortion — not as moral failure, not as laziness, but as temporal misalignment.

Because health is not only about how strong the wave is.

It is about whether it rises at the right time.

And when rhythm falls out of time, restoration begins not with force —

But with re-alignment.

13.1 Insomnia

Insomnia is often described as the inability to sleep.

But from an oscillatory perspective, insomnia is frequently a disorder of **phase** and **recovery**.

The drive for sleep operates through two interacting systems:

1. **Homeostatic sleep pressure** — the gradual buildup of fatigue across the day.
2. **Circadian timing** — the internal clock that determines when sleep is biologically appropriate.

In healthy alignment, these two systems converge at night. Sleep pressure is high. Melatonin rises. Cortisol falls. Neural oscillations slow. Descent occurs naturally.

In insomnia, these systems may be misaligned.

The person may feel tired — but their internal circadian phase remains shifted. Melatonin may not yet be elevated. Cortisol may still be relatively active. The body receives mixed signals.

Attempting sleep during a misaligned phase produces frustration. The more effort applied, the more activation rises. Stress frequency increases. Recovery becomes even more difficult.

Insomnia is not always the absence of fatigue.

It is often sleep attempted at the wrong biological time.

Over time, repeated misalignment can alter:

- **Phase stability**
- **Cortisol rhythms**
- **Immune cycling**
- **Metabolic timing**

The wave still oscillates — but its timing is delayed or fragmented.

Understanding insomnia through phase reduces blame.

The person is not failing to relax.

The rhythm is mistimed.

Restoration often begins with re-anchoring phase — consistent wake time, morning light exposure, reduced evening light, stabilized feeding timing — rather than forcing sleep directly.

Because sleep cannot be commanded.

It must be aligned.

13.2 Shift Work Disorder

Shift work provides a powerful demonstration of phase distortion.

Human physiology evolved under predictable light–dark cycles. The master circadian clock aligns to daylight and darkness. Peripheral clocks in metabolic and immune tissues synchronize accordingly.

When work schedules invert this pattern — waking at night, sleeping during daylight — internal and external cues conflict.

The individual may sleep during the day, but circadian signals still anticipate wakefulness. Cortisol may rise prematurely. Melatonin secretion may truncate. Environmental light intrudes. Sleep depth decreases.

Meanwhile, nighttime activity forces metabolic and cognitive activation during biological night. Insulin sensitivity is reduced. Alertness systems strain against circadian descent.

This is not merely inconvenience.

It is desynchronization at scale.

Shift work disorder often involves:

- Persistent **phase misalignment**
- Impaired **recovery** during daytime sleep
- Elevated inflammatory **baseline**
- Increased metabolic vulnerability

Over time, chronic phase inversion can contribute to cardiovascular risk, mood instability, metabolic syndrome, and immune dysregulation.

The oscillation still exists.

But it runs against environmental time.

Shift work does not imply weakness. Many individuals adapt remarkably well for years. But the underlying biology remains temporally challenged.

Phase distortion under sustained misalignment carries cumulative cost.

Health, in this context, becomes a negotiation between social demands and biological timing.

And the greater the misalignment, the greater the strain on oscillatory integrity.

Because rhythm does not disappear in shift work.

It simply loses its anchor.

13.3 Hormonal Misalignment

Hormones do not circulate randomly.

They follow schedules.

Cortisol rises before waking, preparing the body for activity.
Melatonin rises in darkness, signaling descent into sleep.
Growth hormone pulses during deep sleep.
Insulin sensitivity fluctuates across the day.
Reproductive hormones cycle across weeks and months.

Hormonal health is not simply about levels.

It is about **timing**.

When phase remains aligned, hormonal waves reinforce one another. Cortisol's morning rise supports alertness. Evening melatonin supports sleep. Nighttime growth hormone supports repair. Immune recalibration follows circadian cues.

When phase drifts, hormonal timing unravels.

Cortisol may remain elevated at night.
Melatonin may rise too late or too weakly.
Growth hormone pulses may fragment.
Insulin release may misalign with glucose intake.

The result is not necessarily dramatic endocrine failure.

It is subtle mistiming.

The oscillation still occurs — but at the wrong hour.

Hormonal misalignment can manifest as:

- Difficulty waking despite adequate sleep duration
- Evening alertness that interferes with descent
- Increased nighttime hunger
- Mood variability linked to sleep timing
- Reduced recovery after exertion

Over time, persistent phase distortion can elevate inflammatory tone, narrow emotional amplitude, and weaken metabolic synchronization.

Hormones are oscillators.

When their phase aligns, physiology feels coherent.

When their timing drifts, the body feels slightly out of tune.

13.4 Metabolic Drift

Metabolic drift is rarely abrupt.

It does not arrive overnight.

It unfolds gradually — as small phase distortions accumulate.

Late-night eating shifts peripheral clocks.
Irregular sleep alters cortisol timing.
Light exposure at night suppresses melatonin.
Stress fragments recovery.

Individually, each change may seem minor.

Collectively, they alter **phase** and weaken **synchronization**.

Over time, glucose tolerance declines slightly. Fasting insulin edges upward. Inflammatory baseline rises subtly. Energy becomes less stable.

The wave is still present.

But its alignment degrades.

Metabolic drift reflects phase misalignment compounded by desynchronization. Insulin pulses may occur at suboptimal times. Liver metabolism may anticipate feeding incorrectly. Peripheral clocks may lag behind the central clock.

The body begins to operate with internal disagreement.

This drift does not necessarily produce immediate disease. It creates vulnerability.

The system remains functional — but less resilient.

Metabolic drift illustrates a powerful truth of the Oscillatory Health Model:

Distortion often begins as mistiming.

Health does not collapse all at once.

It gradually loses alignment.

And because drift develops slowly, restoration can also be gradual.

Re-anchoring sleep phase.
Stabilizing meal timing.
Protecting recovery.
Reducing chronic stress frequency.

Phase is not fixed.

It can be recalibrated.

Because rhythm, even when delayed, remembers how to return to time.

Chapter 14 — Desynchronization

When the Orchestra Falls Apart

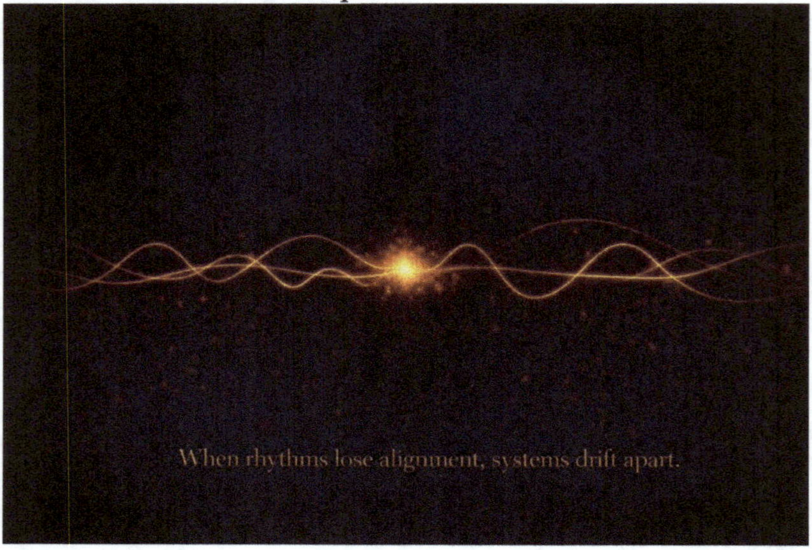

An individual instrument can be perfectly tuned.

The violin may hold pitch.
The percussion may keep steady tempo.
The piano may strike clean notes.

Yet if they do not follow the same conductor, the result is noise.

Desynchronization is not failure of a single rhythm.

It is failure of coordination between rhythms.

The heart may still beat.
Hormones may still pulse.
Glucose may still rise and fall.
Sleep may still occur.

But they no longer align.

In earlier chapters, we examined distortions of **amplitude**, **frequency**, and **phase** within individual systems. Now we turn to something broader:

Loss of **synchronization** across systems.

In healthy physiology, oscillators communicate.

Sleep aligns immune recalibration.
Meal timing coordinates metabolic clocks.
Stress cycles influence inflammatory tone.
Social rhythm stabilizes autonomic balance.

When synchronization weakens, systems begin to operate independently.

The metabolic clock drifts from the circadian clock.
Stress activation persists into sleep phase.
Immune baseline remains elevated during metabolic processing.
Emotional amplitude narrows while stress frequency accelerates.

The body becomes a collection of rhythms without coherence.

Desynchronization often emerges gradually — through chronic stress, sleep disruption, irregular schedules, sustained inflammation, or prolonged emotional strain. No single variable collapses. Instead, communication deteriorates.

The result can manifest as metabolic syndrome, burnout, persistent fatigue, or a vague sense that "everything feels off."

In this chapter, we explore how desynchronization across systems produces complex patterns that cannot be explained by a single variable distortion.

Because sometimes illness is not a broken wave.

It is loss of temporal coordination across interacting systems.

And restoration requires not silencing one instrument —

But restoring timing across them all.

14.1 Metabolic Syndrome

Metabolic syndrome is often described as a cluster of risk factors:

Elevated blood pressure.
Increased abdominal fat.
High fasting glucose.
Altered lipid profiles.

Each marker is measurable. Each has its own biochemical pathway.

But what connects them?

From an oscillatory perspective, metabolic syndrome frequently reflects **desynchronization**.

The central circadian clock may still oscillate. Peripheral clocks in liver and adipose tissue still function. Insulin still pulses. Cortisol still rises in the morning.

Yet the timing between them drifts.

Late-night eating conflicts with melatonin elevation.
Sleep restriction alters cortisol phase.
Chronic stress increases inflammatory baseline.
Irregular schedules fragment synchronization between metabolic tissues.

Over time:

- **Insulin pulses** lose precision.
- **Glucose variability** increases.
- Inflammatory tone subtly rises.
- Energy oscillations flatten.

Metabolic syndrome is not simply excess calories or lack of movement.

It is often the cumulative effect of rhythmic fragmentation.

When feeding, sleep, stress, and light exposure fall out of alignment, metabolic tissues receive mixed timing signals. The orchestra plays — but not together.

Importantly, desynchronization does not produce immediate collapse.

It produces drift.

Years may pass before measurable thresholds are crossed. Yet the oscillatory integrity was weakening long before the diagnosis.

Restoration, therefore, is not only about lowering numbers.

It is about restoring synchronization:

- Stabilizing **phase** through sleep and light.
- Consolidating **feeding windows**.
- Protecting **recovery** after stress.
- Reducing chronic inflammatory **baseline**.

Metabolic syndrome is a visible endpoint of rhythmic misalignment.

It is the body adapting to persistent desynchronization.

14.2 Burnout

Burnout is often framed as exhaustion from overwork.

But burnout is not just fatigue.

It is desynchronization between effort and recovery.

In healthy oscillation, stress activation is followed by descent. Work intensity alternates with restoration. Emotional engagement rises and falls.

In burnout, activation continues — but recovery shortens.

Work demands persist. Emotional labor accumulates. Sleep fragments. Stress frequency accelerates. Immune tone may elevate. Metabolic stability weakens.

The individual may still function. They may continue performing tasks, meeting deadlines, responding to obligations.

But internally:

- **Recovery** becomes incomplete.
- **Baseline stress** rises.
- Emotional **amplitude** narrows.
- Sleep phase destabilizes.

Eventually, motivation dulls. Cynicism increases. Energy flattens.

Burnout is not lack of resilience.

It is oscillation without sufficient synchronization.

Effort remains high while restoration declines.

The system becomes imbalanced not because it cannot rise, but because it cannot descend fully.

Desynchronization between activation and recovery spreads across systems:

Stress cycles invade sleep.
Inflammation rises subtly.
Metabolic timing drifts.
Emotional range compresses.

Burnout is the lived experience of systemic misalignment.

It is the orchestra playing continuously without pause.

Recovery is not indulgence.

It is structural necessity.

When synchronization between effort and restoration is restored — through protected sleep, rhythmic work intervals, social support, and reduced chronic activation — oscillatory integrity can gradually return.

Burnout does not mean the person is broken.

It means the rhythm has been overdriven without adequate descent.

And rhythm, when protected, can be rebuilt.

14.3 Multi-System Fatigue

Multi-system fatigue is different from simple tiredness.

It is not just low energy.

It is the sense that *everything* feels slightly depleted.

The mind struggles to focus.
The body feels heavy.
Motivation dulls.
Sleep does not fully restore.
Stress tolerance decreases.

No single system appears catastrophically impaired. Lab results may hover within normal limits. Yet the overall experience is one of diminished vitality.

From an oscillatory perspective, multi-system fatigue often reflects **desynchronization across multiple variables simultaneously**.

Sleep phase may be slightly delayed.
Stress frequency may be elevated.
Inflammatory baseline may be subtly increased.
Metabolic timing may be irregular.

Individually, each distortion is mild.

Collectively, they fragment coherence.

The result is not dramatic illness.

It is reduced efficiency.

Energy oscillations narrow because recovery windows are insufficient.
Emotional amplitude compresses because stress frequency remains high.
Immune recalibration weakens because sleep depth fragments.

The orchestra is still playing.

But the instruments are no longer reinforcing one another.

Multi-system fatigue represents cumulative rhythmic erosion.

The system continues functioning — but without the dynamic coordination that produces vitality.

This condition often improves not through a single intervention, but through re-synchronization across domains:

- Protecting **sleep phase**
- Stabilizing **feeding windows**
- Reducing chronic **stress frequency**
- Rebuilding **recovery capacity**
- Restoring **social synchronization**

Because fatigue at this scale is not merely energy loss.

It is coherence loss.

14.4 Cross-System Misalignment

Cross-system misalignment occurs when oscillators that once reinforced each other begin to interfere.

Sleep may occur, but at inconsistent times.
Meals may be regular, but at biological night.
Stress activation may persist into recovery windows.
Inflammatory tone may remain elevated during metabolic processing.

Each system retains its own rhythm.

But their timing relationships deteriorate.

For example:

- Elevated evening cortisol delays melatonin.
- Late eating shifts peripheral metabolic clocks.
- Sleep restriction increases inflammatory baseline.
- Chronic inflammation alters mood amplitude.

One distortion cascades into another.

Cross-system misalignment is rarely sudden.

It emerges through accumulated small mismatches.

The danger lies in the feedback loops.

Phase distortion increases stress frequency.
Increased stress frequency impairs recovery.
Impaired recovery elevates baseline inflammation.
Elevated inflammation further destabilizes sleep phase.

The pattern becomes self-reinforcing.

This is why desynchronization can feel complex and difficult to explain. Symptoms cross domains: mood, energy, sleep, metabolism, immunity.

The individual may ask: "What is wrong with me?"

The oscillatory answer is not that something is singularly broken.

It is that the systems are no longer aligned.

Cross-system misalignment illustrates a core principle of this model:

Health is not only about strong individual rhythms.

It is about coordinated timing between them.

When coordination is restored, symptoms across multiple domains often improve together.

Because the goal is not to repair one instrument.

It is to restore the orchestra.

Chapter 15 — Baseline Shift

When the Ground Itself Moves

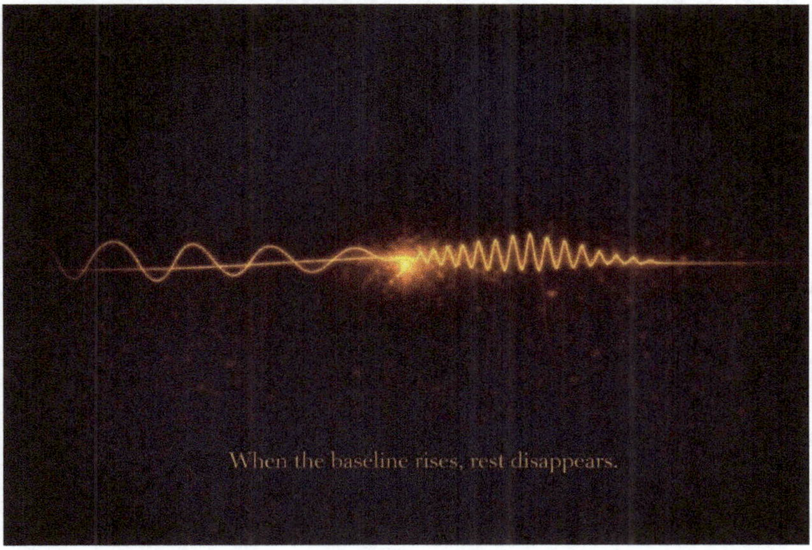

Not all suffering announces itself with extremes.

Sometimes nothing looks dramatic.
There are no visible collapses.
No obvious surges.
No catastrophic spirals.

Instead, something quieter happens.

The floor rises.
Or sinks.
And the body slowly recalibrates to a new normal.

A person who once felt calm now feels slightly tense all the time.
A mind that once rested easily now hovers in low-level vigilance.
Energy never fully returns.
Irritability becomes background noise.
Sleep is lighter.
Recovery is slower.

Nothing is extreme enough to name.
But everything feels subtly altered.

This is a **baseline shift**.

In rhythmic terms, it is not distortion of amplitude or acceleration of frequency.
It is displacement of the set-point around which oscillations occur.

The body still moves.
The mind still adapts.
But it now oscillates around a different center.

Chronic inflammation is not merely prolonged activation — it is immune tone recalibrated upward.
Persistent cortisol elevation is not just stress — it is a new hormonal baseline.
Allostatic load is not momentary strain — it is cumulative reset of regulatory thresholds.

When the baseline shifts, the system no longer returns to where it once did.
It returns to somewhere else.

And over time, that "somewhere else" begins to feel normal.

The danger of baseline shifts is subtlety.
Because nothing dramatic is happening, they are rarely confronted early.
They are tolerated.
Rationalized.
Absorbed.

Until the altered baseline becomes the platform from which future distortions emerge.

In this chapter, we explore what happens when the center moves — biologically, psychologically, and behaviorally.
We examine chronic inflammation, persistent cortisol elevation, and allostatic load not as isolated problems, but as evidence that the regulatory reference point itself has changed.

To understand recovery, we must first recognize when the system is no longer oscillating around health — but around adaptation to strain.

Because sometimes the most important question is not:

"Why am I reacting this way?"

But:

"When did this become my new normal?"

15.1 Chronic Inflammation

When Defense Becomes Atmosphere

Inflammation is not the enemy.

It is one of the body's most intelligent survival tools.
It isolates damage.
Clears debris.
Signals repair.
Mobilizes protection.

In its acute form, inflammation is sharp and purposeful.
It rises.
It acts.
It resolves.

But inflammation is designed to be transient.

When it fails to resolve, something more subtle begins to occur.
The system does not remain in a dramatic storm.
Instead, it settles into a low, persistent hum.

Cytokines circulate at slightly elevated levels.
Microglia remain partially activated.
Immune tone never fully quiets.
The body behaves as if something is always slightly wrong.

This is chronic inflammation — not an explosion, but an atmosphere.

The problem is not intensity.
It is duration.

Low-grade inflammatory signaling changes how cells interpret their environment.
Neural circuits become more reactive.
Energy metabolism shifts toward vigilance rather than restoration.
Stem-cell niches become less permissive.
Repair becomes slower, less precise.

Over time, the immune system recalibrates its baseline.
What was once an emergency signal becomes background noise.

This shift is rarely dramatic enough to diagnose immediately.
There may be fatigue, subtle mood changes, slower recovery from stress, mild cognitive fog, heightened irritability.
Nothing extreme.
But the return-to-rest state no longer feels fully restful.

From a rhythmic perspective, chronic inflammation does not necessarily increase amplitude.
It raises the floor.

Oscillations still occur.
But they now occur around an elevated inflammatory set-point.

This altered baseline has consequences.

When a new stressor arrives, the system begins from an already activated position.
Responses overshoot more easily.
Resolution takes longer.
Minor challenges feel disproportionately exhausting.

The body becomes less flexible — not because it cannot respond, but because it is already partially responding all the time.

Chronic inflammation is therefore not just a disease process.
It is a baseline distortion.

And because it feels normal once established, it often goes unnoticed until larger breakdowns occur.

15.2 Persistent Cortisol Elevation

When Stress Stops Falling Back to Zero

Cortisol is not a villain.

It is a precision hormone.

It mobilizes glucose.
Sharpens attention.
Modulates immune activity.
Coordinates circadian timing.

In healthy oscillation, cortisol rises in the morning, fluctuates with demand, and gradually declines into the night.
It responds to stress — then falls.

But when stress becomes chronic, the rhythm changes.

Instead of clear pulses with recovery, cortisol begins to hover at a higher baseline.
Diurnal variation flattens.
The morning peak may blunt, or the evening level may remain elevated.
The system never fully resets.

This is not the dramatic "fight-or-flight" surge.
It is a sustained shift in hormonal tone.

Persistent cortisol elevation alters multiple regulatory systems simultaneously.

It sensitizes threat detection networks.
It suppresses aspects of immune resolution.
It interferes with deep sleep architecture.
It reshapes hippocampal plasticity.
It biases memory toward vigilance.

Over time, the nervous system begins to treat heightened alertness as normal.

Rest feels unfamiliar.
Stillness feels uneasy.
Calm may even trigger discomfort.

From a baseline perspective, persistent cortisol does not merely increase stress reactivity.
It relocates the body's reference point.

The system oscillates, but around a higher level of anticipatory readiness.

This shift interacts with chronic inflammation in a bidirectional loop.
Elevated cortisol can dysregulate immune signaling.
Chronic inflammatory signals can sustain HPA axis activation.
The two reinforce each other quietly.

Allostasis — the process of adapting to repeated demand — becomes maladaptive when the adaptive set-point never returns.

The body is not failing.

It is adapting continuously.

But adaptation without recovery becomes a new baseline.

And once the baseline shifts, the individual may no longer recognize what unstrained regulation feels like.

They are not constantly stressed.
They are constantly prepared.

And preparation, without pause, slowly becomes exhaustion.

15.3 Allostatic Load

The Cost of Constant Adjustment

The body is not designed for stability.
It is designed for adaptation.

Allostasis means achieving stability through change.
Heart rate rises with demand.
Cortisol increases with threat.
Inflammation activates with injury.
Energy is redirected where it is needed most.

This is not dysfunction.
It is intelligence.

But adaptation has a cost.

When stressors are brief, recovery follows.
The system returns.
Oscillations resume around a stable baseline.

When stressors are repeated, overlapping, or prolonged, recovery becomes incomplete.

The system adapts again.
And again.
And again.

Each adjustment slightly alters regulatory thresholds.
Immune tone shifts upward.
Stress hormones remain partially elevated.

Sleep becomes lighter.
Metabolic efficiency declines.

None of these changes are catastrophic alone.
But together, they accumulate.

This accumulation is called **allostatic load** — the physiological wear and tear produced by repeated adaptation without sufficient restoration.

It is not the stressor itself that causes damage.
It is the lack of return.

Over time, the body becomes efficient at surviving strain but inefficient at resting.

The nervous system anticipates disruption.
The immune system maintains low vigilance.
Endocrine rhythms flatten.

The person may still function.
They may work.
Parent.
Create.
Care.

But the cost of regulation increases.

Small demands feel heavy.
Minor illness lingers.
Emotional recovery slows.
Motivation requires more effort.

Allostatic load does not look like collapse.
It looks like gradual narrowing of resilience.

The system still oscillates — but with reduced flexibility, slower recovery, and higher background tension.

It is a rhythm that still moves, but no longer flows.

15.4 When the Set-Point Moves

The Quiet Redefinition of Normal

Every regulatory system has a reference point.

A temperature around which the body balances.
A hormonal rhythm that anchors the day.
An immune tone that defines safety.
A psychological expectation of calm.

These are set-points — dynamic, not fixed, but centered.

When stress is chronic, inflammation persists, or cortisol remains elevated, the system does something adaptive.

It recalibrates.

Instead of constantly trying to return to an unreachable previous baseline, the body establishes a new one.

Higher vigilance becomes normal.
Lower energy becomes normal.
Mild tension becomes normal.
Interrupted sleep becomes normal.

The person stops noticing the shift because there is no dramatic event marking it.

There is no clear before and after.
Only a gradual relocation of the center.

When the set-point moves, the body still regulates.
It still oscillates.
It still adapts.

But it now oscillates around strain.

This is why recovery can feel confusing.

Someone may say, "I'm not that stressed,"
because acute stress is absent.

Yet they are never fully relaxed.

They may say, "Nothing is wrong,"
because nothing is acutely broken.

Yet they feel subtly depleted.

The difficulty is not that the system cannot regulate.
It is that regulation now serves a different baseline.

Returning to health, then, is not simply reducing symptoms.
It is gently persuading the system that it no longer needs to live at that altered reference point.

Set-points are not moral decisions.
They are physiological compromises.

And like all compromises, they were once protective.

The tragedy is not that the body adapted.

The tragedy is that it adapted for too long.

End of Part III. Disorder Is Often Pattern Distortion, Not Moral Failure

Throughout this part, we have examined rhythm gone awry.

Amplitude flattened.
Amplitude exaggerated.
Frequency accelerated.
Baselines shifted.

None of these distortions arise from weakness.

They arise from systems doing what they were built to do — adapt.

Depression can reflect blunted reward oscillation among other genetic, inflammatory, structural, and developmental influences.
Anxiety can reflect accelerated threat cycling.
Burnout can reflect cumulative allostatic strain.
Chronic stress can reflect a relocated baseline.

These are not character flaws.
They are regulatory patterns under prolonged pressure.

When we mistake pattern distortion for moral failure, we add shame to physiology.
We demand willpower where recalibration is needed.
We prescribe judgment where restoration is required.

Rhythms can distort.
They can narrow.
They can accelerate.
They can shift.

But rhythms can also be retrained.

Healing is rarely about becoming someone new.
It is about restoring flexibility to oscillation.

It is about lowering a baseline gently.
Widening amplitude safely.
Slowing frequency gradually.
Rebuilding the capacity to return.

Disorder is often not a broken self.

It is a body that adapted too well to conditions that lasted too long. And adaptation, once understood, can be guided back toward balance.

Glossary — Part III: Rhythm as Illness

Accelerated Frequency

An increase in the rate at which stress or emotional activation cycles repeat.
Often associated with anxiety, panic loops, hypervigilance, and chronic stress states.

Allostatic Load

The cumulative physiological cost of repeated adaptation without sufficient recovery.
Reflects gradual narrowing of resilience due to persistent regulatory strain.

Baseline Shift

A relocation of the physiological or psychological set-point around which oscillations occur.
May involve chronic inflammation, persistent cortisol elevation, or sustained vigilance.

Blunted Reward Responsiveness

Reduced amplitude of dopamine-mediated reward signaling, often observed in depression and anhedonia.

Burnout

A state of desynchronization between effort and recovery, marked by emotional narrowing, fatigue, and impaired restoration.

Chronic Inflammation

Persistent elevation of immune tone without full resolution.
Represents a raised inflammatory baseline rather than acute immune amplitude.

Chronic Stress Oscillation

Repeated stress activation with incomplete descent, leading to compressed recovery and elevated baseline tension.

Cross-System Misalignment

Loss of coordinated timing between biological systems (sleep, metabolism, immunity, stress), resulting in systemic incoherence.

Desynchronization

Breakdown in temporal coordination between multiple oscillators within the body.
The orchestra continues playing, but not together.

Emotional Narrowing
Reduction in emotional amplitude resulting in muted affect and decreased experiential contrast.

Excess Amplitude
Over-amplified oscillation in mood, energy, or stress responses.
Seen in bipolar spectrum conditions and emotional volatility.

Flattened Amplitude
Reduced height of emotional, motivational, or energetic oscillations.
Often present in depression, anhedonia, and chronic fatigue.

Hypervigilance
Persistently accelerated low-level stress frequency characterized by continuous environmental scanning and incomplete recovery.

Metabolic Drift
Gradual deterioration of metabolic synchronization due to accumulated phase distortion and desynchronization.

Metabolic Syndrome (Oscillatory Lens)
A cluster of metabolic risk factors reflecting long-term rhythmic misalignment across feeding, sleep, stress, and inflammatory systems.

Oscillation Without Control
High-amplitude cycling lacking sufficient regulatory containment, leading to destabilizing peaks and troughs.

Panic Loop
A rapid self-amplifying cycle of physiological arousal and threat interpretation that escalates frequency and amplitude simultaneously.

Persistent Cortisol Elevation
Chronic upward shift in cortisol baseline with reduced diurnal variation and impaired recovery.

Phase Delay
Misalignment of internal biological timing relative to environmental cues, commonly seen in insomnia and shift work disorder.

Set-Point Relocation
The adaptive but maladaptive recalibration of physiological baseline in response to chronic strain.

Stress Spike
A disproportionately large amplitude stress response relative to the triggering stimulus.

Temporal Distortion
A mismatch in the timing, pacing, or spacing of biological oscillations.

Trauma-Linked Frequency Acceleration
Sustained increase in stress cycle repetition following prolonged threat exposure.

Wave Distortion
Any deviation in amplitude, frequency, phase, synchronization, recovery, or baseline that reduces oscillatory flexibility.

Pattern Distortion
The central thesis of Part III:Many disorders can be understood, at least in part, as rigid or maladaptive alterations in oscillatory variables rather than moral or personal failure.

PART IV — Restoring Rhythm

How to Reclaim the Wave

Rhythms distort.
But they also recalibrate.

The body is not fragile. It is adaptive.
What becomes rigid can soften.
What becomes chaotic can stabilize.
What becomes flat can rise again.

Restoration is not the elimination of fluctuation.
It is the strengthening of modulation.

Not suppression.
Not force.
Not perfection.

Modulation.

In this section, we move deliberately through each variable — not to repeat theory, and not to offer rigid prescriptions — but to identify leverage points within the oscillatory system.

When you can name the variable under strain — **amplitude, frequency, phase, synchronization, recovery, or baseline** — intervention becomes specific instead of overwhelming.

Distortion feels global.
Repair is precise.

Rhythms are plastic.
Plasticity does not imply unlimited self-correction. Some conditions require medical treatment, structured psychotherapy, pharmacologic stabilization, or specialized care. The strategies that follow are not replacements for appropriate clinical intervention. They are complementary levers — ways to strengthen modulation within safe and responsible boundaries.

The nervous system adapts.
Hormonal cycles recalibrate.
Circadian clocks re-anchor.
Immune tone stabilizes.
Emotional range expands.

The question is not whether the wave will move.
It will.

The question is whether it will move with coherence.

Restoring rhythm does not mean flattening intensity or eliminating stress.
It means strengthening the relationship between rise and return.

Rebuilding amplitude without volatility.
Regulating frequency without suppression.
Anchoring phase without rigidity.
Re-synchronizing systems through shared cues.
Protecting recovery.
Stabilizing baseline without dulling vitality.

This section offers a practical architecture for rhythmic repair.

Not control of every symptom.
But influence over timing, pacing, and descent.

Agency does not come from eliminating vulnerability.
It comes from understanding where flexibility can be rebuilt.

Health is rarely unshakeability.
It is coherence under movement.

The wave will continue to rise and fall.
The work now is to help it rise with strength —
and return with stability.

Tool 4.1 — Identify Your Primary Oscillatory Distortion

Before attempting to restore rhythm, it helps to identify which variable may be under the greatest strain.

This is not a diagnostic instrument.
It is a structural reflection tool.

Consider your experience over the past several weeks or months.
Which pattern feels most familiar?

Amplitude — Height of Rise and Fall

- Do your emotions feel muted or flattened?
- Do positive experiences feel less intense than they once did?
- Or do you experience extreme emotional highs and lows?

If so, your **amplitude** may be reduced or excessive.

Frequency — Speed of Cycles

- Do stress responses repeat throughout the day?
- Does your mind cycle rapidly through worry or rumination?
- Do you feel constantly "on," without pause?

If so, your **frequency** may be accelerated.
Or:

- Do you feel slowed, dull, or stuck in prolonged states?

If so, your **frequency** may be sluggish.

Phase — Timing Alignment

- Is your sleep schedule delayed or unstable?
- Do you feel most alert at night and tired during the day?
- Does your energy feel mistimed relative to your responsibilities?

If so, **phase misalignment** may be present.

Synchronization — System Coordination

- Do your sleep, mood, energy, and appetite feel out of sync?
- Does physical fatigue coexist with mental restlessness?
- Do different areas of your life feel rhythmically disconnected?

If so, **desynchronization** may be involved.

Recovery — Return to Baseline

- After stress, do you remain activated for hours?
- Do minor conflicts linger internally?
- Do you struggle to "come down" after stimulation?

If so, **recovery** may be incomplete.

Baseline — Resting Set-Point

- Do you feel persistently tense, inflamed, or on edge even without triggers?
- Does low mood feel like your default state?
- Does your body rarely feel fully at ease?

If so, your **baseline** may be elevated or depressed.

Interpreting Your Pattern

Most people will recognize more than one variable.

However, one often feels primary.

Restoration begins not by correcting everything at once, but by targeting the dominant distortion first.

If amplitude is low → expand safely.
If frequency is high → slow deliberately.
If phase is delayed → anchor timing.
If synchronization is fragmented → align systems.
If recovery is incomplete → protect descent.
If baseline is elevated → reduce chronic activation.

The chapters that follow provide structured approaches for each pathway.

Chapter 16 — Rebuilding Amplitude

Teaching the Wave to Rise Again

When amplitude narrows, life becomes quiet.

Flattened amplitude often presents as diminished motivational drive, reduced affective contrast, lower behavioral initiation, and attenuated reward anticipation.

Earlier, we explored how flattened **amplitude** can manifest as depression, anhedonia, chronic fatigue, or emotional narrowing. The system adapts by dampening peaks — sometimes to protect itself from overwhelm, sometimes in response to prolonged stress or inflammation.

But amplitude is not meant to remain small.

Amplitude gives vitality.

It allows joy to feel vivid.
Effort to feel purposeful.
Sadness to feel meaningful.
Engagement to feel alive.

The question is not how to force intensity.

The question is how to *rebuild amplitude safely*.

Rebuilding amplitude requires paradox.

To strengthen the rise, you must protect the return.

To increase intensity, you must preserve recovery.

Amplitude without modulation leads to volatility. But amplitude paired with strong **recovery** produces resilience.

This chapter explores how amplitude can be expanded through:

- **Physical stress paired with recovery**, teaching the body to tolerate stronger activation.
- **Emotional range expansion**, allowing feelings to deepen without overwhelming the system.
- **Behavioral variability**, introducing novelty and contrast into daily life.
- **Controlled intensity**, practicing strong activation within safe boundaries.

Amplitude is like a muscle.

Underused, it weakens.
Overdriven, it destabilizes.
Trained progressively, it strengthens.

Rebuilding amplitude does not mean chasing constant highs.

It means restoring healthy contrast.

Because vitality does not come from eliminating the trough.

It comes from trusting that the peak can rise — and still return.

Amplitude is not a luxury.

It is the height of the wave that makes life vivid.

And when rebuilt with care, it becomes a source of strength rather than instability.

16.1 Physical Stress + Recovery

The body rebuilds amplitude through contrast.

One of the most reliable ways to restore healthy **amplitude** is through controlled physical stress — paired deliberately with recovery.

Exercise is not merely calorie expenditure. It is oscillatory training.

When you lift weight, sprint, or challenge endurance, you intentionally increase physiological amplitude:

- Heart rate rises.
- Stress hormones pulse.
- Muscles strain.
- Energy demand surges.

This is the upward arc.

But the true benefit emerges in the descent.

During recovery:

- Parasympathetic tone increases.
- Inflammation resolves adaptively.
- Growth hormone supports repair.
- Baseline strengthens rather than drifts upward.

Without recovery, physical stress narrows amplitude over time. Overtraining leads to fatigue, mood instability, and baseline elevation.

With recovery, amplitude expands.

The system learns it can rise strongly and return safely.

This principle extends beyond elite athletics. Even moderate, rhythmic stress — brisk walking, resistance training, interval training, cold exposure, heat exposure — can gently widen amplitude when balanced with sleep, nourishment, and rest.

The key is pairing activation with restoration.

Stress without recovery destabilizes.
Recovery without stress weakens contrast.

Together, they train the wave.

16.2 Emotional Range Expansion

Just as muscles can atrophy, emotional amplitude can narrow.

If sadness is avoided, joy flattens.
If anger is suppressed, clarity dulls.
If vulnerability is denied, connection thins.

Emotional range expansion is the intentional widening of amplitude — not through chaos, but through awareness and gradual exposure.

This may include:

- Allowing sadness to be felt fully without self-judgment.
- Engaging in activities that provoke genuine excitement.
- Having difficult but honest conversations.
- Practicing creative expression.

Emotional expansion does not require dramatics.

It requires permission.

When emotion is experienced without immediate suppression, the nervous system learns that amplitude is tolerable. The rise does not destroy stability. The fall does not erase safety.

Importantly, expansion must be paired with recovery.

After intense emotion, rest.
After vulnerability, integration.
After challenge, grounding.

The goal is not to live at high amplitude.

It is to preserve range.

Emotional health depends not on constant positivity, but on flexibility — the capacity to move across a wide arc and return.

Range restores vitality.

And vitality depends on amplitude that is strong, safe, and supported.

16.3 Behavioral Variability

Amplitude narrows when life becomes overly predictable.

The same routines.
The same environments.
The same conversations.
The same cognitive loops.

Predictability stabilizes **baseline**, but too much uniformity compresses contrast. Without variation, the nervous system receives limited stimulus for dynamic rise.

Behavioral variability restores contrast safely.

This does not require dramatic life changes. It requires intentional shifts in pattern:

- Trying new physical activities.
- Learning unfamiliar skills.
- Changing social environments.
- Altering daily routines in small ways.
- Introducing creative exploration.

Novelty increases dopaminergic engagement. Curiosity expands motivational amplitude. The brain responds to contrast.

Importantly, variability must be tolerable.

Overwhelming novelty destabilizes.
Gradual variability strengthens.

Think of behavioral variability as widening the experiential bandwidth of daily life. Small changes reintroduce rise into flattened patterns. The system learns to expand without losing recovery.

Amplitude thrives on contrast.

Behavioral monotony shrinks it.

Intentional variability rebuilds it.

16.4 Controlled Intensity

Intensity is not the problem.

Unregulated intensity is.

Controlled intensity is the deliberate practice of high amplitude within boundaries that protect **recovery**.

This may include:

- High-intensity interval training with adequate rest.
- Deep creative immersion followed by detachment.
- Emotional vulnerability within safe relationships.
- Public speaking with grounding afterward.
- Focused work sprints with structured breaks.

The pattern is consistent:

Strong rise.
Clear containment.
Protected descent.

Controlled intensity teaches the nervous system that high activation does not require collapse. Peaks become integrated rather than destabilizing.

Without control, intensity can trigger excessive amplitude, sleep disruption, or prolonged stress frequency. With control, intensity expands capacity.

This is rhythmic training.

You are not eliminating strong waves. You are strengthening your ability to ride them.

Over time, the system becomes more confident in its ability to rise and return. Emotional peaks feel less threatening. Physical stress feels less overwhelming. Motivation can surge without spiraling.

Amplitude is not restored by avoidance. It is restored by safe expansion.

Controlled intensity rebuilds the height of the wave — without sacrificing stability.

And when amplitude returns with modulation intact, vitality follows.

Chapter 17 — Regulating Frequency

Slowing the Tempo

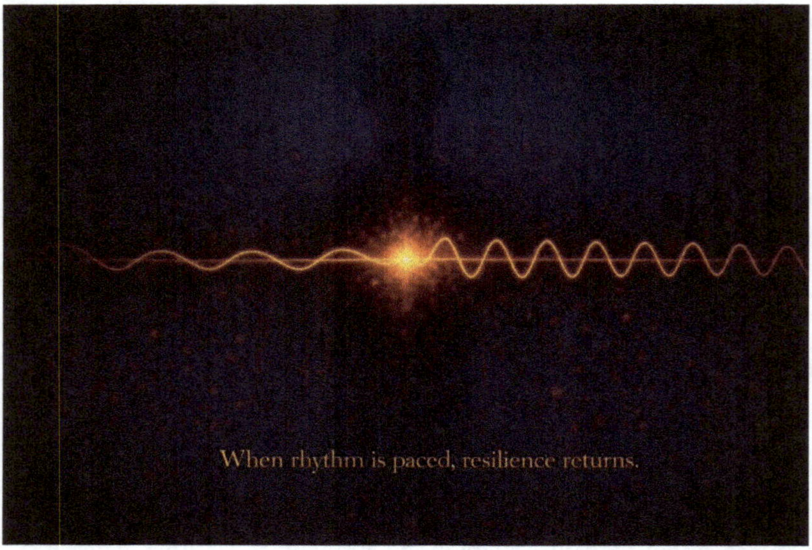

Some systems suffer not because the wave is too small.

Not because it is too high.

But because it never stops coming.

When frequency accelerates, activation cycles begin to overlap. Cognitive engagement, physiological arousal, and stress signaling occur in rapid succession, leaving insufficient spacing for recovery.

There is no space between pulses.

When **frequency** accelerates, the nervous system loses margin. Recovery shortens. Baseline tension rises. Even mild stressors feel amplified — not because they are stronger, but because they arrive too quickly.

Earlier, we explored how anxiety cycles, panic loops, hypervigilance, and chronic stress reflect increased frequency. The body remains in repeated activation without sufficient descent.

Now we ask a different question:

How do we slow the tempo?

Regulating frequency does not mean eliminating activation. It means creating spacing between cycles.

Spacing allows:

- Complete **recovery**
- Restoration of **baseline**
- Protection of **amplitude**
- Preservation of **synchronization**

When pulses are spaced appropriately, the system regains flexibility.

This chapter explores practical mechanisms for regulating frequency:

- **Breath pacing**, which directly influences autonomic oscillation.
- **Attention cycles**, which restore cognitive spacing.
- **Stress interval design**, which structures activation and recovery intentionally.
- **Nervous system cadence**, the broader rhythm of daily activation and rest.

Frequency is not fixed.

It responds to pacing.

Just as music becomes chaotic when played too fast, physiology destabilizes when cycles accelerate without control.

Slowing does not weaken the system.

It strengthens it.

Because resilience requires not only strong waves —

But space between them.

17.1 Breath Pacing

Breath is the only autonomic function we can consciously modulate at will.

Heart rate, digestion, inflammatory tone — these respond indirectly. Breath, however, sits at the intersection of voluntary and automatic control.

This makes it a powerful regulator of **frequency**.

When stress accelerates, breathing shortens and quickens. Shallow, rapid breaths signal activation. The nervous system interprets this pacing as urgency.

When breathing slows and deepens, the signal reverses.

Longer exhales stimulate the **vagus nerve**, increasing parasympathetic tone. Heart rate variability improves. Cortisol frequency reduces. Recovery spacing widens.

This shift can be observed through improved beat-to-beat heart rate variability — one of the most widely studied markers of autonomic regulation.

Breath pacing works not because it eliminates stress, but because it alters temporal pacing.

For example:

- Inhaling for 4 seconds
- Exhaling for 6 seconds
- Sustaining this pattern for several minutes

This shifts the rhythm of internal oscillators. Activation pulses slow. The system regains temporal margin.

In clinical and experimental settings, paced breathing at approximately six breaths per minute has been shown to increase heart rate variability — a measurable indicator of autonomic flexibility — within minutes. Over time, consistent practice can widen recovery windows, reduce stress frequency, and strengthen parasympathetic modulation without requiring extreme intervention.

Breath pacing is not about hyperventilation or extreme control. It is about gentle entrainment.

The goal is coherence.

Regular breath pacing trains the nervous system to:

- Slow activation frequency
- Strengthen recovery descent
- Protect baseline stability

When practiced consistently, breath becomes a metronome.

And the nervous system follows its tempo.

17.2 Attention Cycles

The mind oscillates too.

Focus rises, sustains, then declines. Cognitive energy is not infinite. Yet modern environments encourage continuous engagement — scrolling, multitasking, prolonged concentration without pause.

This accelerates cognitive **frequency**.

Thought loops tighten. Mental fatigue increases. Stress activation overlaps with attention effort.

Healthy attention follows cycles:

Focus → Release → Focus → Release.

Short work intervals followed by deliberate pauses restore spacing. Even brief disengagement — standing, walking, closing the eyes, shifting environment — widens the gap between activation pulses.

Structured attention cycles may include:

- 25–50 minutes of focused work
- 5–10 minutes of complete disengagement
- Periodic longer breaks across the day

This pattern prevents cognitive frequency from compressing.

Without spacing, attention becomes continuous activation. With spacing, the mind regains rhythm.

Attention cycles protect:

- Emotional amplitude
- Stress recovery
- Sleep phase integrity
- Baseline cognitive stability

Regulating frequency is not about doing less.

It is about pacing.

When focus and rest alternate intentionally, mental oscillation remains adaptive rather than frantic.

The nervous system does not need constant stillness.

It needs cadence.

And cadence restores resilience.

17.3 Stress Interval Design

Stress cannot — and should not — be eliminated.

The goal is not a stress-free life.

The goal is rhythmic stress.

Stress interval design is the intentional structuring of activation and recovery across the day.

Instead of allowing stress to accumulate unpredictably, you organize it into defined intervals:

Activation → Containment → Recovery → Baseline.

For example:

- A focused work sprint followed by deliberate rest.
- A difficult conversation followed by grounding and decompression.
- Physical exertion paired with hydration and stillness.

When stress is continuous and unstructured, **frequency** compresses. Activation bleeds into sleep. Recovery fragments. Baseline tension rises.

When stress is interval-based, the system learns something critical:

Peaks are temporary.

This predictability reduces anticipatory hypervigilance. The nervous system does not remain on standby because it trusts that recovery is coming.

Interval design strengthens:

- **Recovery efficiency**
- Emotional modulation
- Sleep phase stability
- Baseline resilience

It transforms stress from chaos into rhythm.

The body does not suffer from activation alone.

It suffers from activation without scheduled descent.

17.4 Nervous System Cadence

Cadence is the larger rhythm of a life.

It is not just breath cycles or work intervals.

It is the tempo of days and weeks.

Some lives operate at relentless speed — constant urgency, constant input, constant stimulation. Others swing between overdrive and collapse.

Healthy nervous system cadence balances activation and restoration across multiple timescales:

Daily cadence:
Morning rise → Midday peak → Evening descent → Night recovery.

Weekly cadence:
Effort days → Lighter days → Social connection → Rest.

Seasonal cadence:
Periods of growth → Periods of consolidation.

When cadence collapses into constant acceleration, frequency distortion spreads across systems. Sleep shortens. Immune tone elevates. Emotional amplitude narrows.

When cadence becomes stagnant, amplitude weakens.

Balanced cadence protects oscillatory integrity.

It allows strong peaks without chronic acceleration.

Practical cadence stabilization may include:

- Consistent sleep and wake times.
- Structured work–rest cycles.
- Technology boundaries.
- Predictable social engagement.
- Intentional recovery rituals.

Cadence is not rigidity.

It is rhythmic scaffolding.

When the nervous system experiences predictable alternation between effort and restoration, it slows naturally. Frequency stabilizes. Recovery deepens.

You cannot eliminate waves.

But you can shape their tempo.

And when tempo is restored, resilience strengthens.

Because health requires not constant calm —

But sustainable rhythm.

Chapter 18 — Anchoring Phase

Returning the Wave to Time

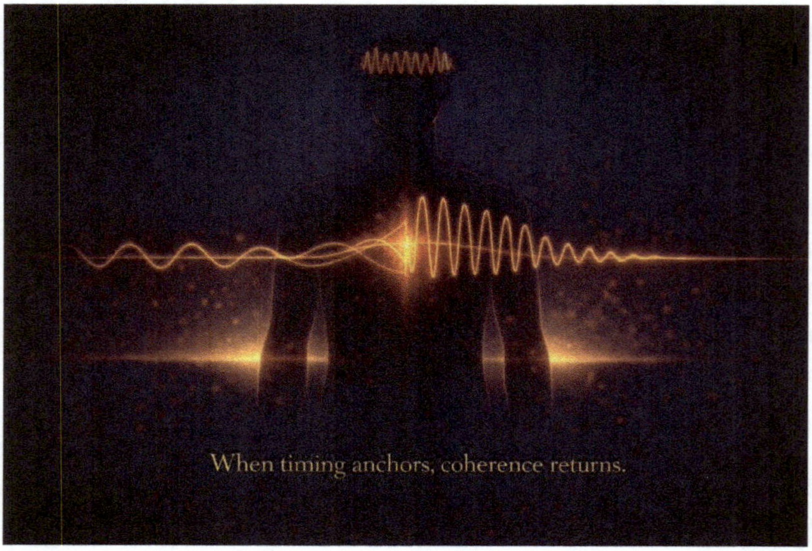

You can have strong amplitude.

You can regulate frequency.

You can protect recovery.

And still feel off.

Not broken. Not exhausted. Just slightly misaligned.

Alert at midnight.
Foggy at 8 a.m.
Hungry late at night.
Unmotivated in the afternoon.

The wave is present.

But it is out of time.

Phase is the variable of alignment.

It determines *when* activation occurs — not just how much. It anchors internal oscillators to environmental cues, especially light and darkness.

Without a stable phase anchor, other variables destabilize.

Recovery may occur at the wrong hour.
Insulin pulses may misalign with feeding.
Cortisol may rise too late.
Melatonin may be suppressed too long.

The orchestra may still be playing.

But without a conductor, the timing drifts.

Modern life exerts constant pressure on phase:

Artificial light after sunset.
Irregular sleep schedules.
Late-night stimulation.
Travel across time zones.
Shift work.

The circadian system adapts — but not without cost.

Anchoring phase is not about discipline.

It is about coherence.

This chapter explores how **light timing**, **sleep windows**, **meal timing**, and applied **chronobiology** can stabilize the internal clock and realign oscillators across systems.

When phase is anchored:

- Hormones rise at appropriate times.
- Energy aligns with daylight.
- Recovery deepens at night.
- Metabolic rhythms strengthen.

The goal is not perfection.

It is predictability.

Because when the wave knows when to rise —

Everything else becomes easier.

Phase is the spine of rhythm.

And when it is anchored, the body remembers its timing.

18.1 Light Timing

Light is the most powerful **phase anchor** available to the human body.

It does not merely help us see.

It tells us when to rise and when to descend.

Specialized cells in the retina detect brightness and transmit that information directly to the brain's master circadian clock. Morning light signals activation. Evening darkness signals restoration.

When light timing is consistent, internal oscillators align:

- **Cortisol** rises shortly after waking.
- **Melatonin** declines in the morning and rises predictably at night.
- Alertness increases during daylight.
- Sleep pressure aligns with darkness.

When light exposure is mistimed — bright light late at night, dim light in the morning — phase drifts.

The brain interprets late-night brightness as extended daytime. Melatonin is suppressed. Sleep onset delays. Cortisol timing shifts. Recovery compresses.

Morning light is especially powerful.

Exposure to natural daylight within the first hour of waking strengthens phase stability. Even 10–30 minutes outdoors can reinforce alignment. Cloudy days still provide sufficient brightness to anchor the clock.

Evening darkness is equally important.

Reducing artificial light after sunset, dimming indoor lighting, and limiting screen exposure help preserve melatonin timing.

Light timing does not require perfection.

It requires contrast.

Bright days.
Dark nights.

This contrast anchors phase and strengthens synchronization across metabolic, immune, and emotional systems.

Light is not a lifestyle preference.

It is biological instruction.

18.2 Sleep Windows

Sleep does not occur randomly.

There is a biologically optimal window when the body is most prepared for descent.

This window reflects alignment between **circadian phase** and accumulated sleep pressure.

When sleep occurs consistently within a stable window — even if exact duration varies slightly — phase coherence strengthens. Hormonal timing stabilizes. Recovery deepens. Morning activation improves.

Irregular sleep timing fragments phase.

Sleeping at 10 p.m. one night and 1:30 a.m. the next sends mixed signals to the circadian clock. The internal rhythm struggles to predict descent. Cortisol timing may shift. Melatonin release may become inconsistent.

A sleep window does not require rigid bedtimes.

It requires consistency within a predictable range.

For example:

- Sleeping between 10:30 p.m. and 11:30 p.m. most nights.
- Waking at a similar time daily — even on weekends.

The wake time is particularly important. Consistent waking reinforces morning light exposure and stabilizes cortisol rise.

Sleep windows protect:

- **Phase alignment**
- **Recovery efficiency**
- Emotional amplitude
- Metabolic timing

When sleep timing stabilizes, other oscillators begin to synchronize naturally.

Sleep is not only about duration.

It is about timing.

And when the window is protected, the wave descends and rises in harmony with the day.

18.3 Meal Timing

Food is more than fuel.

It is a timing signal.

Every time you eat, peripheral clocks in the liver, pancreas, gut, and adipose tissue adjust. These tissues respond not only to nutrients, but to when nutrients arrive.

When meals occur at predictable times — ideally during daylight hours — metabolic clocks synchronize with the central circadian clock in the brain.

Insulin pulses align with glucose intake.
Digestive enzymes activate in anticipation.
Energy rises and falls predictably.

When meals are irregular — skipped mornings, late-night eating, inconsistent schedules — peripheral clocks drift. The body receives conflicting cues.

Light may signal night.
Food may signal day.

This internal disagreement weakens **phase alignment** and reduces **synchronization**.

Meal timing does not require extreme restriction. It requires coherence.

- Eating within a consistent daily window.
- Avoiding heavy meals during biological night.
- Aligning the largest meal earlier in the day when insulin sensitivity is higher.

These practices strengthen phase stability across metabolic tissues.

When feeding aligns with daylight and sleep, energy stabilizes more naturally. Recovery improves. Baseline inflammatory tone may decrease.

The body thrives on predictable nourishment.

Timing is a metabolic regulator.

18.4 Chronobiology in Daily Life

Chronobiology is the science of biological timing.

But in practice, it is about rhythm literacy.

It is the awareness that nearly every physiological process follows a temporal pattern — and that daily behavior can either reinforce or disrupt those patterns.

Applying chronobiology in daily life means:

- Exposing yourself to bright light in the morning.
- Dimming lights in the evening.
- Protecting consistent sleep windows.
- Aligning meals with daylight hours.
- Scheduling demanding cognitive work during natural alertness peaks.
- Reserving evenings for descent rather than stimulation.

It does not require perfection.

It requires awareness.

Modern environments often fragment time cues. Artificial lighting blurs day and night. Digital devices compress rest. Work schedules override natural rhythms.

Chronobiology restores contrast.

Day for activation.
Night for recovery.

When these boundaries strengthen, internal oscillators regain coherence.

Phase anchoring does not solve every problem. But it provides structure upon which other variables stabilize.

Without anchored phase, regulating amplitude and frequency becomes more difficult.

With anchored phase, recovery deepens, synchronization improves, and baseline stabilizes.

The wave remembers its timing.

And when timing returns, vitality often follows naturally.

Chapter 19 — Re-Synchronizing Systems

Bringing the Orchestra Back Together

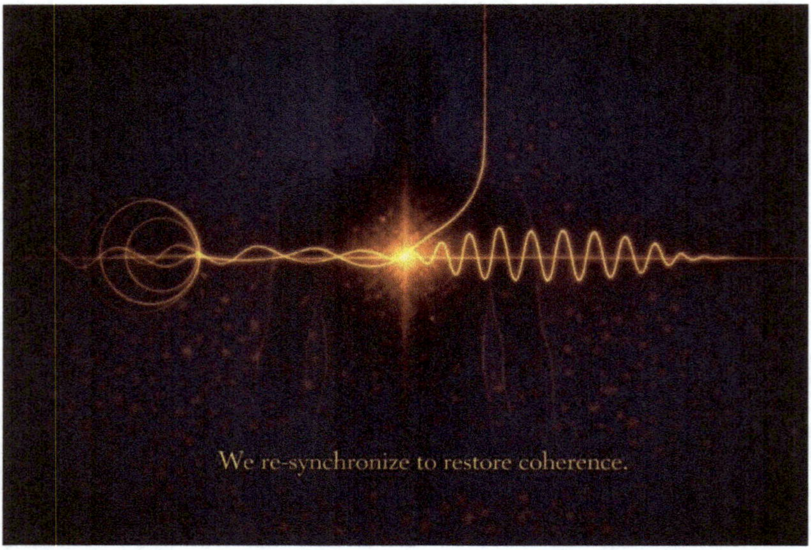

You can strengthen amplitude.
You can slow frequency.
You can anchor phase.

And still feel fragmented.

Because health is not only about individual variables.

It is about how they align.

Sleep may improve, yet metabolism feels unstable.
Stress may decrease, yet inflammation remains elevated.
Energy may rise, yet mood feels flat.

These patterns often reflect incomplete **synchronization**.

Earlier, we explored desynchronization as illness — when oscillators drift apart. Now we focus on restoration.

Re-synchronization is not about correcting one system in isolation.

It is about creating signals that multiple systems can follow simultaneously.

Exercise timed appropriately can align metabolic and circadian clocks.
Anti-inflammatory behaviors can quiet immune baseline and stabilize mood.
Consistent social rhythms can anchor autonomic oscillation.
Environmental cues can entrain sleep, stress, and energy together.

The body does not heal through isolated adjustments alone.

It heals through coordinated signaling.

Think of synchronization as relational timing within the organism. When sleep, metabolism, immunity, emotion, and stress follow compatible rhythms, they reinforce each other.

When they fall out of step, energy leaks across systems.

This chapter explores practical strategies for restoring coherence:

- **Exercise timing** as a cross-system synchronizer.
- **Anti-inflammatory behavior** to stabilize baseline.
- **Social rhythm therapy** to regulate autonomic timing.
- **Environmental entrainment** to strengthen circadian alignment.

Synchronization is not perfection.

It is coherence.

The orchestra does not need silence.

It needs alignment.

And when systems rise and fall together — even imperfectly — vitality returns with less effort.

Because the body is not designed to function as isolated parts.

It is designed to move as one coordinated rhythm.

19.1 Exercise Timing

Exercise is not only a tool for building strength.

It is a **synchronizing signal**.

When timed intentionally, physical activity can align multiple oscillators at once — metabolic, circadian, immune, and emotional systems.

Morning exercise, especially in natural light, reinforces **phase anchoring**. It amplifies the natural rise of cortisol, increases alertness, and strengthens circadian alignment.

Midday exercise can enhance metabolic responsiveness, improve insulin sensitivity, and stabilize energy amplitude.

Evening exercise, if intense and too close to sleep, may delay descent by elevating core temperature and sympathetic activation. But moderate movement earlier in the evening can help discharge stress and improve nighttime recovery.

Timing matters.

Exercise acts as a cue that tells the body:

"This is active time."

When consistently placed within the day, it strengthens rhythmic predictability. Heart rate variability improves. Sleep depth deepens. Inflammatory tone may decrease over time.

Exercise is one of the few interventions that:

- Expands **amplitude**
- Strengthens **recovery**
- Stabilizes **baseline inflammation**
- Reinforces **phase alignment**

Its synchronizing power lies not only in intensity, but in regularity.

Movement, when rhythmic and predictable, becomes a conductor.

19.2 Anti-Inflammatory Behavior

Inflammation is not inherently harmful.

But elevated inflammatory **baseline** can destabilize mood, energy, and metabolic regulation.

Anti-inflammatory behavior does not mean suppressing the immune system. It means restoring healthy pulsatility and preventing chronic elevation.

Key behaviors that support immune synchronization include:

- Consistent, sufficient sleep
- Whole-food nutrition rich in fiber and phytonutrients

- Regular moderate exercise
- Stress frequency regulation
- Social connection

These behaviors reduce persistent inflammatory signaling and protect recovery cycles.

Importantly, inflammation is sensitive to desynchronization. Sleep disruption elevates cytokines. Chronic stress accelerates immune frequency. Late-night eating misaligns metabolic and immune clocks.

Even modest elevations in inflammatory markers such as C-reactive protein or interleukin-6 have been associated with measurable changes in mood regulation, energy balance, and cognitive flexibility.

When baseline inflammation stabilizes:

- Emotional amplitude becomes more flexible
- Stress recovery improves
- Metabolic timing strengthens
- Sleep deepens

Anti-inflammatory behavior is not a diet trend.

It is rhythmic hygiene.

The immune system functions best when activation is episodic and resolution is complete.

Restoring immune pulsatility strengthens synchronization across systems.

Because when baseline inflammation lowers, other oscillators regain freedom to move.

And the orchestra begins to sound coherent again.

19.3 Social Rhythm Therapy

Human beings do not regulate alone.

Our daily routines — when we wake, eat, work, speak, rest, and connect — create rhythmic scaffolding for the nervous system. When those routines fragment, oscillators drift.

Social Rhythm Therapy emerged from the recognition that mood stability is closely tied to regular daily patterns. While originally developed in the context of mood disorders, its principles extend broadly to oscillatory health.

The core idea is simple:

Stabilize the timing of daily activities.

- Wake at a consistent hour.
- Eat meals at predictable times.
- Schedule social interaction regularly.
- Anchor work and rest cycles.

These routines provide external timing cues that reinforce **phase alignment** and cross-system **synchronization**.

When daily rhythms are chaotic, internal rhythms follow.

Sleep shifts.
Stress frequency increases.
Emotional amplitude destabilizes.

When daily rhythms stabilize, the body regains predictability.

Social rhythm is not rigidity.

It is patterned repetition that protects modulation.

Even small consistencies — a morning walk at the same time, a shared evening meal, a weekly gathering — strengthen autonomic coherence.

Connection reinforces timing.

And timing restores synchronization.

19.4 Environmental Entrainment

Entrainment is the process by which oscillators align with external cues.

Light entrains circadian phase.
Temperature cycles influence sleep onset.
Soundscapes shape autonomic tone.
Natural environments regulate stress frequency.

Modern environments often fragment entrainment signals. Artificial light blurs night. Noise disrupts descent. Temperature remains constant across seasons. Screens compress rest and work into the same sensory field.

Environmental entrainment restores contrast.

- Bright natural light during the day.
- Darkness and reduced stimulation at night.
- Cooler temperatures for sleep.
- Natural sound exposure when possible.
- Clear boundaries between work and recovery spaces.

The body listens to the environment continuously.

When cues are coherent, oscillators align more easily. When cues conflict, phase and synchronization weaken.

You do not need perfect surroundings.

You need intentional cues.

Environment is silent instruction.

When it reinforces day–night contrast, recovery deepens. When it supports calm in the evening, frequency slows. When it exposes the body to predictable patterns, baseline stabilizes.

Re-synchronization does not always begin inside.

Sometimes it begins by changing the signals around you.

Because the orchestra responds not only to internal timing —

But to the stage on which it plays.

Chapter 20 — Meaning as Long-Wave Rhythm

The Slow Oscillations That Hold a Life Together

Up to now, we have spoken about waves measured in seconds, minutes, hours, and days.

Earlier chapters focused on oscillations measured in seconds, minutes, and days — cardiac variability, respiratory pacing, sleep architecture, endocrine pulses, and stress intervals.

But not all rhythms are fast.

Some move slowly.

So slowly that we mistake them for personality.
For destiny.
For identity.

This chapter returns to where *The Happiness Rebound* began — with the arc of meaning.

If dopamine pulses represent short-wave amplitude, meaning represents a **long-wave oscillation**. It does not rise and fall within minutes. It unfolds across years.

There are seasons of expansion.
Seasons of contraction.
Seasons of building.
Seasons of grief.

Meaning is not constant elevation.

It is patterned movement across a life.

When long-wave rhythm destabilizes, everything feels disoriented — even if sleep is stable and stress is manageable. Without coherent purpose, emotional amplitude narrows. Frequency accelerates in search of stimulation. Baseline drifts.

Meaning stabilizes the system not by eliminating fluctuation, but by contextualizing it.

A difficult season becomes a phase.
A failure becomes a trough in a larger arc.
A transition becomes part of a cycle rather than an ending.

The nervous system responds differently when experience fits into a coherent narrative.

Identity provides continuity across oscillation.

Purpose provides direction through descent.

Culture provides shared timing.

Narrative integrates peaks and troughs into pattern.

This final chapter expands the Oscillatory Health Model beyond physiology.

Because health is not only the regulation of hormones and neurons.

It is the capacity to endure rise and fall without losing coherence.

Short waves give us vitality.

Long waves give us meaning.

And when both align, the organism feels whole.

We now turn to the slow oscillations — the rhythms measured not in heartbeats, but in chapters of a life.

Because ultimately,

Health is not the elimination of cycles.

It is the ability to move through them — and return.

20.1 Purpose as Slow Oscillation

Purpose is not constant intensity.

It is not perpetual motivation.
It is not uninterrupted clarity.

Purpose behaves like a slow oscillation.

There are seasons when it rises — when direction feels clear, effort feels meaningful, and energy aligns with intention. Projects move forward. Identity feels expansive. Momentum builds.

And there are seasons when it descends.

Motivation softens. Doubt increases. Energy narrows. The future feels less defined.

In *The Happiness Rebound*, we explored how emotional peaks inevitably fall — and how that descent is not failure but recalibration. The same principle applies at the scale of purpose.

Long-wave oscillation includes:

- Expansion → Consolidation
- Exploration → Commitment
- Effort → Reflection
- Growth → Integration

When individuals interpret descent as collapse rather than phase, distress increases. They attempt to force constant ascent — constant productivity, constant passion, constant visibility.

But long-wave health depends on accepting oscillation across years.

Purpose does not require permanent elevation.

It requires directional continuity through fluctuation.

When purpose functions as a slow oscillation, it stabilizes faster variables:

- Emotional amplitude becomes contextualized.
- Stress frequency feels purposeful rather than chaotic.
- Recovery becomes part of growth rather than interruption.

Purpose does not eliminate difficulty.

It integrates it.

It transforms troughs into preparation rather than failure.

The wave of meaning rises and falls slowly — but its arc gives coherence to the shorter waves beneath it.

20.2 Identity Stability

If purpose is the long-wave movement of direction, identity is the baseline that holds it.

Identity stability does not mean rigidity.

It means continuity across oscillation.

You can experience emotional highs and lows without losing your sense of self.
You can shift careers without losing identity.
You can enter seasons of doubt without dissolving coherence.

In oscillatory language, identity functions like a stabilized **baseline** at the level of narrative.

When baseline identity is unstable, every emotional fluctuation feels existential. A mistake becomes proof of inadequacy. A conflict becomes threat to selfhood. A setback feels like collapse.

When identity is stable, fluctuations are interpreted as part of the arc.

Identity stability allows:

- Emotional amplitude without self-fragmentation.
- Stress activation without identity threat.
- Purpose shifts without total disorientation.

This does not mean clinging to a fixed persona. It means maintaining an adaptable core — values, principles, narrative continuity.

In rhythmic terms:

Amplitude can rise and fall.
Frequency can accelerate and slow.
Phase can shift.

But baseline identity remains coherent.

Identity is not a static label.

It is a dynamic continuity.

When identity holds, oscillation feels survivable.

And when oscillation feels survivable, resilience strengthens.

Purpose gives direction to the wave.

Identity gives it ground.

Together, they create long-wave stability that supports every shorter rhythm in the body.

20.3 Cultural Rhythm

No life oscillates in isolation.

We rise and fall within larger collective rhythms — families, communities, nations, traditions.

Culture has cadence.

Workweeks and weekends.
Religious observances.
Seasonal holidays.
Academic years.
Harvest cycles.
Fiscal quarters.

These shared structures create predictable long-wave oscillations that scaffold individual timing.

When cultural rhythm is coherent, it stabilizes identity and purpose. Shared rituals reinforce belonging. Collective pauses create synchronized recovery. Seasonal shifts normalize cycles of expansion and contraction.

But modern culture often accelerates.

Continuous connectivity.
Blurring of work and rest.

Global time zones collapsing day and night.
Endless information flow.

Cultural cadence compresses.

Without collective pauses, individuals must self-regulate against constant activation. Without shared rituals, meaning becomes individualized and fragile.

Cultural rhythm provides temporal containment.

It says:

"This is a season of work."
"This is a season of rest."
"This is a season of celebration."
"This is a season of grief."

When culture respects oscillation, individuals feel less pressure to maintain constant elevation.

The long wave of society shapes the long wave of the self.

Health is not only biological alignment.

It is rhythmic participation.

20.4 Narrative Integration

The final restoration is not physiological.

It is narrative.

Narrative integration is the ability to interpret one's life as patterned movement rather than random disruption.

Without narrative, oscillation feels chaotic.

With narrative, oscillation becomes meaningful.

A depressive episode becomes a trough in a longer arc.
A burnout phase becomes a signal of desynchronization.
A period of intense ambition becomes an amplitude expansion.
A loss becomes descent before renewal.

Narrative does not deny suffering.

It contextualizes it.

In *The Happiness Rebound*, we explored how emotional descent is built into the architecture of joy. In this book, we have extended that principle across physiology.

Narrative integration unifies them.

You are not broken because you experience cycles.

You are oscillatory by design.

When individuals can identify which variable has distorted — amplitude, frequency, phase, synchronization, recovery, or baseline — they regain agency. They move from confusion to pattern recognition.

And pattern recognition restores dignity.

Narrative is the long-wave integration of rhythm into identity.

It allows the organism to endure rise and fall without interpreting descent as failure.

It transforms oscillation into coherence.

Chapter 21 — Putting It Together

How to Identify Your Pattern

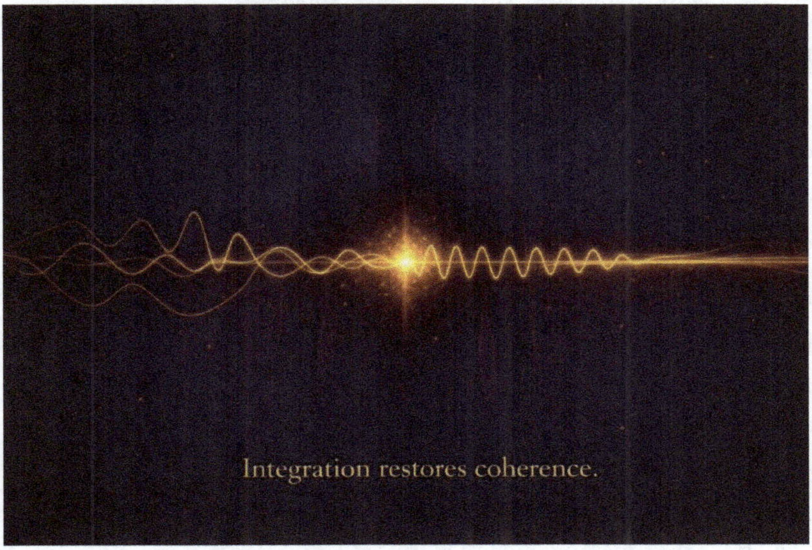

Understanding variables is powerful.
Applying them is transformative.

Throughout this book, we have traced oscillation from molecule to mood, from sleep to immunity, from light to metabolism. We have seen that health is not a fixed state but a living movement — amplitude rising and falling, phase aligning and drifting, recovery restoring what activation disrupts.

We have learned the grammar of rhythm.

But grammar alone does not restore coherence.

Architecture explains.
Application heals.

Until now, the model has lived on the page — in heart rate variability, cortisol waves, immune pulses, emotional arcs, circadian clocks. Now it must live in you.

Because the Oscillatory Health Model is not theoretical.
It is experiential.

It appears in the hour you wake at night.
In how long tension lingers after conflict.
In whether joy feels vivid or muted.
In whether stress resolves — or accumulates.
In whether your energy rises cleanly — or fragments.

The question shifts here.

No longer:

"What is the model?"

But:

"How is my rhythm behaving?"

Symptoms can feel chaotic.
But patterns have shape.

This chapter is not about diagnosing disease.
It is about recognizing structure.

We will walk through a composite case.
We will identify which variables distort.
We will trace the order in which they shifted.
We will design a modulation sequence.

And in doing so, you will begin to see something quietly powerful:

You do not have five unrelated problems.

You have one wave that drifted.

Once you see the pattern, intervention becomes strategic instead of reactive.

Let us walk through an example.

A Composite Case

Forty-two.

Not ill.
Not in crisis.
Functioning.

But something has shifted.

Sleep fractures.

Waking at 3–4 a.m., several nights each week.
Not panicked. Not startled.
Just awake — as though descent no longer completes.

Afternoon energy collapses without warning.
Focus narrows.
Sugar calls louder than it used to.

Irritability lingers longer than it once did.
Small frustrations echo through the evening.

Weight gathers slowly at the abdomen,
despite similar meals, similar routines.

And at night — when fatigue should invite rest —
the body feels tired,
but the mind hums.

Wired.
But tired.

Nothing catastrophic.
No diagnosis.
No trauma.
No single moment to point toward.

Just subtle erosion.

It feels like several small problems at once:

Sleep.
Energy.
Mood.
Metabolism.

Separate systems.
Separate explanations.

But we have learned something different.

Instead of asking,

"What is wrong?"

We ask:

Which variables are distorted?

Step 1 — Identify the Dominant Variables

The symptoms feel scattered.

But patterns are not random.

We begin by asking:

Which variables are under strain?

Phase

Waking at 3–4 a.m.
Late-night alertness.

These suggest phase misalignment —
the internal clock drifting subtly out of alignment with light and darkness.

The rhythm remains — but its timing has drifted.

Recovery

Irritability that lingers.
Difficulty settling after stimulation.

These suggest impaired recovery —
activation rising appropriately,
but descent failing to complete.

The wave climbs.
But it does not fully return.

Baseline

Persistent tension.
Gradual abdominal weight gain.
Energy that never quite resets.

These suggest a mild elevation of baseline —
a resting state that has crept upward.

Not acute stress.
Not overt inflammation.

Just chronic, low-level activation.

And already something becomes clear.

This is not six separate problems.

It is three interacting distortions:

Phase.
Recovery.
Baseline.

The architecture is beginning to reveal itself.

Step 2 — Determine What Shifted First

Patterns distort in sequence.

And sequence matters.

We ask:

Which variable moved earliest?

Often, the first shift is subtle.

In this case, bedtime gradually drifted later over several years.
Screens extended into the night.
Morning light exposure shortened.

The change was incremental.
Barely noticeable.

Phase likely shifted first.

And from there, the cascade began:

Phase delay
→ fragmented sleep

Fragmented sleep
→ impaired recovery

Impaired recovery
→ gradual baseline elevation

One small misalignment.
Then another.
Then another.

Distortion spreads not because the body fails —
but because oscillators are coupled.

This is why sequence matters.

We do not attempt to correct everything at once.

We correct the origin.

Step 3 — Design the Modulation Sequence

Once the origin becomes visible, intervention becomes orderly.

We do not scatter effort.

We restore sequence.

Anchor Phase

If phase drifted first, phase must be corrected first.

Not dramatically.
Not aggressively.
But consistently.

- Morning light exposure within the first hour of waking
- A stable wake time — even after imperfect nights
- Reduced late-night light and screen intensity

Light anchors the clock.
Consistency stabilizes the arc.

When phase realigns, downstream oscillators begin to follow.

Protect Recovery

With sleep timing stabilizing, we protect descent.

Recovery is not passive.
It is structured.

- A defined wind-down period before bed
- Emotional closure — journaling, reflection, or resolution of unfinished loops
- Slow breathing or parasympathetic activation before sleep

The goal is not sedation.

The goal is completion.

Activation must have an ending.

When recovery deepens, irritability shortens.
Stress cycles regain curvature.

Lower Baseline

Only after phase and recovery begin to stabilize do we address baseline.

• Moderate exercise earlier in the day
• Reduced evening stimulation
• Anti-inflammatory behavioral patterns — consistent meals, regular movement, restorative sleep

Baseline rarely drops through force.

It lowers when cycles complete.

Notice what we did not do:
We did not eliminate stress.
We did not chase perfect sleep.
We did not attack weight directly.
We restored modulation.

Step 4 — Observe Systemic Effects

Restoration is rarely immediate.
But it is cumulative.

After four to six weeks:

• Sleep consolidates
• Irritability resolves more quickly
• Afternoon crashes soften
• Abdominal weight stabilizes
• Subjective tension decreases

Not because each symptom was individually targeted.

But because rhythm was recalibrated.

When phase aligns, recovery deepens.
When recovery deepens, baseline lowers.
When baseline lowers, amplitude returns.

The system remembers its shape.

What This Demonstrates

Symptoms rarely travel alone.

They cluster around variables.

What appears to be multiple failures is often one distortion expressed across systems.

Variables do not collapse simultaneously.

They distort in sequence.

Phase shifts.
Recovery shortens.
Baseline drifts.

In another pattern, amplitude may flatten first.
Or frequency may accelerate before phase shifts.

And because oscillators are coupled, one subtle change spreads.

Restoration also follows sequence.

Correct the origin.
Support the cascade.
Allow downstream systems to recalibrate.

Targeting the primary distortion reduces overwhelm.

Instead of attacking five symptoms,
we adjust one variable.

Instead of suppressing expression,
we restore structure.

This is where the power of the framework lies:

It converts complexity into architecture.

And architecture can be repaired.

Apply This to Yourself

You do not need laboratory data to begin.

You need observation.

Ask:

Which variable feels most strained?

Is it:

- **Amplitude** — does life feel flattened, or excessively reactive?
- **Frequency** — does everything feel too fast, or too slow?
- **Phase** — is sleep mistimed? Is energy misaligned with the day?
- **Synchronization** — do systems feel out of step with one another?

• **Recovery** — does activation linger long after it should resolve?
• **Baseline** — is there constant tension, or persistent fatigue, even at rest?

Then ask:

When did this begin?
What likely shifted first?
What was the earliest subtle drift?

And most importantly:

What intervention restores modulation
rather than suppresses symptoms?

Do not aim for total correction.

Do not attempt to fix everything at once.

Aim for the first lever.

Small realignments propagate.

Rhythms influence rhythms.

When one variable stabilizes, others often follow.

Because the body does not heal by force.

It heals through coordinated return.

Important Clarification

Not every pattern can be self-modulated.

Some distortions require:

• Medical evaluation
• Pharmacologic stabilization
• Structured psychotherapy
• Specialist care

There are moments when amplitude is too narrow to rise on its own.
When baseline is too elevated to settle without intervention.
When phase disruption reflects deeper pathology.
When recovery has been impaired for too long.

This framework does not replace treatment.

It does not compete with medicine.

It enhances structural understanding.

It allows you to see pattern beneath symptom.

It allows participation in modulation within appropriate boundaries.

Even when professional care is necessary, rhythm still matters.

Medication may stabilize amplitude.
Therapy may strengthen recovery.
Medical intervention may lower baseline.

But the architecture remains.

And understanding architecture restores agency.

The Larger Insight

Illness feels chaotic when we look only at symptoms.

Poor sleep.
Low mood.
Fatigue.
Inflammation.
Irritability.

They appear separate.
Unrelated.
Overwhelming.

But patterns feel manageable because they have shape.

The Oscillatory Health Model does not reduce human complexity.

It organizes it.

It does not simplify experience into numbers.

It reveals structure beneath fluctuation.

You do not have six unrelated failures.

You have one distorted wave.

And waves, when understood,

can be guided back into motion.

End of Part IV. Health Is Rhythmic Integrity, Not Constant Control

We began this book with a simple thesis:

Health is not static stability — it is flexible biological oscillation.

Across sleep, mood, immunity, metabolism, stress, and connection, we have traced the hidden architecture beneath vitality. We have seen that life unfolds not in straight lines, but in waves — rising, falling, synchronizing, recovering.

We have examined how illness often begins quietly — not as sudden collapse, but as **distortion of rhythm**.

A narrowing of **amplitude**.
An acceleration of **frequency**.
A drift in **phase**.
A fracture in **synchronization**.
An incomplete **recovery**.
A subtle shift in **baseline**.

Small distortions accumulate.
Flexibility narrows.
The system stiffens before it breaks.

Modern culture often teaches us to pursue **control** — to flatten emotion, suppress inflammation, override fatigue, silence stress, stabilize everything at once.

But the body was never designed for constant control.

It was designed for **modulation**.

Health is not the absence of fluctuation.
It is not perpetual elevation.
It is not rigid calm.

Health is **rhythmic integrity**.

It is the preservation of adaptive **amplitude** — the ability to rise fully and descend cleanly.
It is sustainable **frequency** — activation with space to recover.
It is aligned **phase** — internal timing anchored to light and environment.

It is cross-system **synchronization** — rhythms moving in coordinated harmony.
It is complete **recovery** — return to rest after activation.
It is a flexible **baseline** — a resting tone that does not drift into chronic elevation or collapse.

When these variables remain fluid, the organism remains resilient.

When they narrow, fragment, or desynchronize, vitality fades — not because life has failed, but because its oscillations have lost coherence.

Illness is often not chaos.
It is **rigidity**.

Health is not perfection.
It is the capacity to move through cycles and return.

To activate — and recover.
To expand — and contract.
To experience stress — and rebound.
To feel joy — and recalibrate.
To lose alignment — and re-anchor.

The body does not require stillness.

It requires movement within bounds.

It requires waves that can rise and fall without fragmentation.

At every scale — molecular, neural, hormonal, immune, social — life expresses this same principle:

Oscillation sustains coherence.

And when oscillation remains flexible, healing remains possible.

Health is not constant control.

It is the integrity of the wave.

Glossary — Part IV: Restoring Rhythm

Anti-Inflammatory Behavior
Daily practices that stabilize immune baseline and restore immune pulsatility without suppressing necessary activation. Includes sleep protection, whole-food nutrition, stress regulation, and rhythmic exercise.

Behavioral Variability
Intentional introduction of novelty and contrast into daily routines to rebuild adaptive amplitude and prevent emotional flattening.

Breath Pacing
Conscious regulation of inhalation and exhalation timing to slow autonomic frequency, increase vagal tone, and widen recovery intervals.

Cadence
The broader tempo of daily, weekly, and seasonal activation–recovery cycles. Healthy cadence balances effort and restoration across time scales.

Chronobiology
The study and application of biological timing principles in daily life, including light exposure, sleep timing, and meal alignment.

Controlled Intensity
Deliberate engagement in high-amplitude activation within structured boundaries that protect recovery and prevent volatility.

Environmental Entrainment
Alignment of internal rhythms with external cues such as light, temperature, sound, and spatial boundaries to strengthen phase and synchronization.

Exercise Timing
Strategic placement of physical activity within the day to reinforce circadian alignment, metabolic responsiveness, and cross-system synchronization.

Frequency Regulation
The intentional slowing or spacing of activation cycles to restore recovery integrity and reduce baseline elevation.

Light Timing
Exposure to bright light during the day and darkness at night to anchor circadian phase and stabilize hormonal oscillation.

Modulation
The central therapeutic principle of Part IV.
Adjusting amplitude, frequency, phase, synchronization, recovery, or baseline without eliminating natural fluctuation.

Nervous System Cadence
The patterned alternation between activation and restoration across daily life that preserves oscillatory flexibility.

Oscillatory Plasticity
The capacity of biological rhythms to recalibrate through repeated behavioral cues and environmental alignment.

Phase Anchoring
Stabilizing internal biological timing through consistent light exposure, sleep windows, and predictable routines.

Recovery Protection
Intentional safeguarding of descent phases following activation to prevent baseline drift and cumulative strain.

Re-Synchronization
The restoration of coordinated timing between multiple biological systems through shared cues (movement, sleep, social rhythm, environment).

Rhythmic Hygiene
Daily behaviors that preserve oscillatory integrity across systems (consistent sleep, aligned meals, paced stress, environmental contrast).

Social Rhythm Therapy
Stabilization of daily activity timing (wake time, meals, work, social interaction) to reinforce phase alignment and mood stability.

Stress Interval Design
Structuring activation and recovery into predictable cycles rather than allowing continuous, uncontained stress exposure.

Timing Hygiene
Preservation of consistent environmental and behavioral time cues to strengthen circadian and metabolic coherence.

Wave Training

The practice of intentionally strengthening oscillatory capacity through controlled activation paired with protected return.

Long-Wave Meaning
Slow oscillations of purpose, identity, and narrative that provide coherence across shorter biological rhythms.

Rhythmic Integrity
The state in which amplitude, frequency, phase, synchronization, recovery, and baseline are flexibly modulated rather than rigid or distorted.

EPILOGUE — The Grace of Return

Where the Wave Comes Home

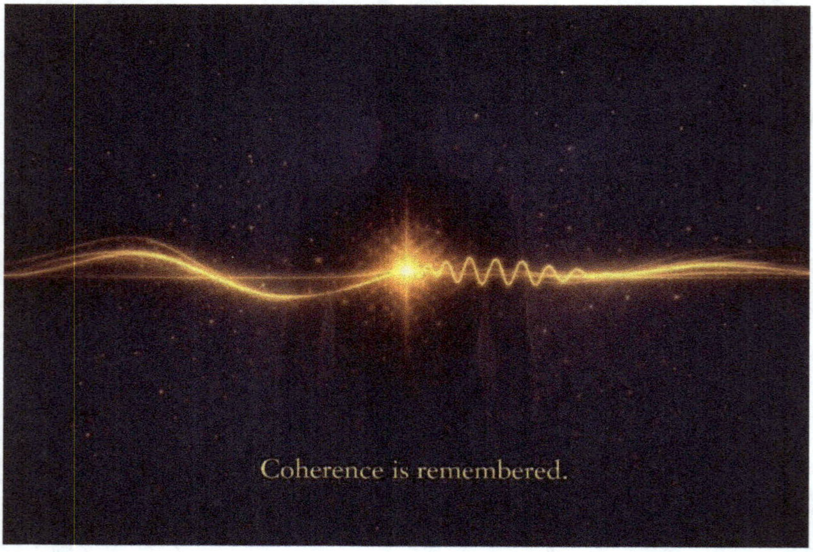

When the body steadies after strain, there is no announcement.
No visible line marking the moment of recalibration.

There is only a quiet shift —
breath deepening,
shoulders lowering,
the nervous system remembering its **range**.

Modern life quietly teaches us that health means never falling.
Never getting tired.
Never feeling low.
Never losing focus.
Never drifting.

But biology does not seek permanence.
It seeks **rhythm**.

The heart does not hold a single beat.
The lungs do not sustain one breath.
Hormones do not remain elevated without consequence.

Emotion does not stay high without descent.
Immunity does not activate without resolving.
Metabolism does not mobilize without recalibrating.

Life is **patterned movement**.

Distortion is rarely dramatic.
It is incremental.

A shortened **recovery**.
An accelerated **frequency** of stress.
A narrowing of **amplitude**.
A drifting **phase**.
A subtle shift in **baseline**.
A quiet loss of **synchronization**.

The wave continues to move —
but with less coherence.

And yet —
distortion is not destiny.

Because rhythm is **plastic**.

Given steady cues —
light anchored,
sleep protected,
breath paced,
stress completed,
connection restored —

oscillators begin to realign.

Not instantly.
Not dramatically.
But gradually.

Variability returns.
Timing stabilizes.
Range expands.
Energy regains contour.
Mood regains depth.

Immunity regains resolution.
Metabolism regains rhythm.

Restoration is not a breakthrough.
It is repetition.

Return to light.
Return to sleep.
Return to pacing.
Return to recovery.
Return to connection.
Return to meaning.

Health does not demand constant elevation.
It requires **return**.

Return after stress.
Return after drift.
Return after excess.
Return after loss.

Not perfection — but **modulation**.

The wave does not need to stay high.
It needs to keep moving.

Because health is not flawless control.
It is **rhythmic integrity under change**.

The capacity to rise.
To fall.
To recalibrate.
And to come home again.

And somewhere — perhaps at 3:17 a.m. —
when sleep deepens,
when cortisol falls,
when the immune pulse quiets,
when breath lengthens into the dark —

the rhythm returns.

Always, the wave remains possible.
Not frozen at its peak —

but rising, descending,
and rising again.

Further Readings

Circadian timing, phase, and whole-body coordination

- **Chill & Wright (2024)** — *Eating around the clock: circadian rhythms of eating and metabolism.*
- **Reytor-González et al. (2025)** — *Chrononutrition and Energy Balance: How Meal Timing and Circadian Rhythms Shape Weight Regulation and Metabolic Health.*
- **Ma et al. (2025)** — *Time-restricted eating and circadian rhythms: A new frontier in diabetes and obesity management.*
- **Chen et al. (2026)** — *Effects of timing and eating duration of time restricted eating on metabolic outcomes: systematic review and network meta-analysis.*

Sleep as master oscillator and immune recalibration

- **Singh et al. (2024)** — *Sleep and Immune System Crosstalk: Implications for Inflammatory Homeostasis and Disease Pathogenesis.*
- **Li (2025)** — *Development of a circadian immune system.*

Brain clearance and sleep physiology (glymphatic / CSF dynamics)

- **Dagum et al. (2026)** — *The glymphatic system clears amyloid beta and tau from brain to plasma in humans.*
- **Tang et al. (2025)** — *Glymphatic Function Alterations in Sleep Disorder Patients Post-COVID-19: A Longitudinal DTI-ALPS Study.*
- **Berberich et al. (2025)** — *Increased coupling between global grey matter and CSF-derived fMRI signal in young adults after partial sleep deprivation – evidence from the sleepy brain study: Increased gGM-CSF-Coupling after sleep deprivation.*

Mood, light, and circadian disruption

- **Dollish et al. (2023/2024)** — *Circadian Rhythms and Mood Disorders: Time to See the Light.*
- **Deprato et al. (2025)** — *Associations between light at night and mental health: A systematic review and meta-analysis.*
- **Lee et al. (2024)** — *The real-world association between digital markers of circadian disruption and mental health risks.*

Autonomic rhythm, HRV, and breath pacing (frequency regulation)

- **Schneider et al. (2025)** — *Associations Between Positive Affect and Heart Rate Variability: A Systematic Review.*
- **Zhao et al. (2025)** — *Heart rate variability and its modulation by nutrients: a narrative review on implications for cardiovascular aging.*
- **Dolgilevica et al. (2025)** — *Heart Rate Variability Biofeedback Training Can Improve Menopausal Symptoms and Psychological Well-Being in Women with a Diagnosis of Primary Breast Cancer: A Longitudinal Randomized Controlled Trial.*

Social synchronization and rhythm therapy

- **Sankar et al. (2021)** — *Telehealth Social Rhythm Therapy to Reduce Mood Symptoms and Suicide Risk Among Adolescents and Young Adults With Bipolar Disorder.*
- **Darwish et al. (2024)** — *Effect of combining social rhythm therapy to treatment of bipolar disorder versus treatment as usual: a comparative study on a sample of Egyptian patients.*

About the Author

Moawiah Naffaa, PhD, writes at the intersection of neuroscience, physiology, and lived human experience. His work is guided by a quiet but foundational question beneath modern life: what patterns shape our internal worlds?

In *Rhythm as Medicine, Rhythm as Illness*, he reframes health not as a static condition, but as a living oscillation — a coordinated movement across sleep, mood, metabolism, immunity, stress, and social connection. Drawing from contemporary neuroscience, chronobiology, psychophysiology, and systems theory, he introduces the Oscillatory Health Model, a framework for understanding how biological timing sustains vitality and how its distortion gives rise to illness.

Dr. Naffaa's writing brings together scientific rigor and narrative clarity. Rather than treating bodily systems as separate domains, he illuminates their hidden synchrony — how cortisol interacts with immune signaling, how sleep recalibrates inflammation, how emotional range reflects neurological flexibility, and how resilience depends less on suppression than on rhythmic integrity.

He invites readers to look beyond diagnosis and beneath fragmentation — to see health as coordinated architecture rather than isolated malfunction.

Through science, structure, and story, his work seeks to reconnect readers with the quiet intelligence of their own biology — and to remind them that life does not sustain a single note.

It breathes between them.

www.ingramcontent.com/pod-product-compliance
Lightning Source LLC
LaVergne TN
LVHW010719061126
851857LV00014B/1260